Optimization Methods for User Admissions and Radio Resource Allocation for Multicasting over High Altitude Platforms

RIVER PUBLISHERS SERIES IN COMMUNICATIONS

Indexing: All books published in this series are submitted to the Web of Science Book Citation Index (BkCI), to CrossRef and to Google Scholar.

The "River Publishers Series in Communications" is a series of comprehensive academic and professional books which focus on communication and network systems. Topics range from the theory and use of systems involving all terminals, computers, and information processors to wired and wireless networks and network layouts, protocols, architectures, and implementations. Also covered are developments stemming from new market demands in systems, products, and technologies such as personal communications services, multimedia systems, enterprise networks, and optical communications.

The series includes research monographs, edited volumes, handbooks and textbooks, providing professionals, researchers, educators, and advanced students in the field with an invaluable insight into the latest research and developments.

For a list of other books in this series, visit www.riverpublishers.com

Optimization Methods for User Admissions and Radio Resource Allocation for Multicasting over High Altitude Platforms

Ahmed Ibrahim

Memorial University of Newfoundland
Canada

Attahiru S. Alfa

University of Pretoria
South Africa

and

University of Manitoba
Canada

River Publishers

Published, sold and distributed by:
River Publishers
Alsbjergvej 10
9260 Gistrup
Denmark

River Publishers
Lange Geer 44
2611 PW Delft
The Netherlands

Tel.: +45369953197
www.riverpublishers.com

ISBN: 978-87-7022-036-1 (Hardback)
       978-87-7022-035-4 (Ebook)

Contents

Preface

This book is based on part of the PhD research conducted by Ahmed Ibrahim [1] under the supervision of Prof. Attahiru Alfa, for optimizing *radio resource allocation* (RRA) and user *admission control* (AC) for multiple multicasting sessions on a single *high-altitude platform* (HAP) with multiple antennas on-board. HAPs are quasi-stationary aerial platforms that carry a wireless communications payload to provide wireless communications and broadband services. They are meant to be located in the stratosphere layer of the atmosphere at altitudes in the range 17–22 km and have the ability to fly on demand to temporarily or permanently serve regions with unavailable telecommunications infrastructure.

An important requirement that we focus on in this book is the development of an efficient and effective method for resource allocation and user admissions for HAPs, especially when it comes to multicasting. Power, frequency, space (antennas selection) and time (scheduling) are the resources considered in the problem. The combination of these many aspects of the problem in multicasting over an *orthogonal frequency division multiple access* (OFDMA) HAP system were not, to the best of our knowledge, addressed before we conducted the research in this area. Due to the strong dependence of the total number of users that could join different multicast groups, on the possible ways we may allocate resources to these groups, it is of significant importance to consider a joint user to session assignments and RRA across the groups. From the service provider's point of view, it would be in its best interest to be able to admit as many higher priority users as possible, while satisfying their quality of service requirements. High-priority users could be users subscribed in and paying higher for a service plan that gives them preference of admittance to receive more multicast transmissions, compared to those paying for a lower service plan. Also, the user who tries to join multiple multicast groups (i.e. receive more than one multicast transmission), would have preferences for which one he would favor to receive if resources are not enough to satisfy the quality of service (QoS) requirements of all the sessions it desires to receive.

A mathematical optimization problem is formulated, which turns out to be a mixed integer non-linear non-convex program for which branch and bound solution framework is guaranteed to solve the problem. Branch and bound can be also used to obtain sub-optimal solutions with known goodness. Even though branch and bound is guaranteed to find the optimal solution, the computational cost could be too high. In this book, we focus on how to enhance the performance of the branch and bound algorithm such that the computational effort and the tree size (which affects the memory requirements) are reduced. Two key things we rely on in this book for improving computational performance are reformulations and relaxations. In reformulations, we reduce the problem into good structured problems for which efficient algorithms have been developed in *operations research* (OR) literature, which we use with some modification to suit our problem. Different relaxations are used to obtain good bounds that are used for early pruning of nodes in the branch and bound tree.

In this book, we extend earlier published works on this type of problem of RRA over HAPs [2–4]. The extended problem that we take into account in this book allows the multicasting group users to receive a session's transmission from more than one antenna simultaneously at different frequencies. It also allows the user to receive multicast sessions transmitted in neighboring cells too, not just those transmitted in the cell in which the user resides. Moreover, in this book, different users have different priorities in the system and for each user, every multicast session requested could have different levels of importance for that user. The objective is to maximize the admissions of the highest-priority users to the system, with priority for each user to be admitted first to the sessions of more importance.

An important contribution in this book is deriving a formulation for the extended problem which is much more efficient, in terms of size, as compared to the one derived in our earlier published works [2–4]. For the derived formulation of the extended problem in this book, based on its observed features, our proposed solution method uses linear outer approximation with McCormick under-estimators for the relaxation of the mixed binary quadratically constrained problem formulation. The proposed solution method is based on branch-and-cut scheme in which cutting planes, domain propagation and different types of heuristics are integrated. Various branching schemes are considered in this book, and a presolving reformulation linearization scheme for a specific set of quadratic constraints in the derived formulation is performed and discussed in this book.

The numerical experiments compare the performances in terms of the duality gap, number of nodes, number of iterations, the number of iterations per node, the time needed to obtain the first feasible solution and the percentage of instances a feasible solution was found.

List of Figures

List of Tables

List of Abbreviations

AC	Admission control
AC-RRA	Joint admission control and radio resource allocation
AVG	Average
BER	Bit-error-rate
BnB	Branch and Bound
BS	Base Station
CDMA	Code Division Multiple Access
CPU	Central Processing Unit
CSI	Channel State Information
E-SysMod	Extended System Model
FDMA	Frequency Division Multiple Access
GEO	Geostationary Earth Orbit
HAP	High Altitude Platform
IDT	Information Decomposition Techniques
ITU	International Telecommunications Institute
ITU-R	International Telecommunications Institute-Radio communications
LCG	Least Channel Gain
LEO	Low Earth Orbit
L.H.S	Left Hand Side
LOS	Line-of-sight
MBLP	Mixed Binary Linear Program
MBPC	Mixed Binary Polynomial Constraint
MBPCP	Mixed Binary Polynomial Constrained Program
MBQCP	Mixed Binary Quadratically Constrained Program
MINLP	Mixed Integer Non Linear Program
MIQCP	Mixed Integer Quadratically Constrained Program
OFDM	Orthogonal Frequency Division Multiplexing
OFDMA	Orthogonal Frequency Division Multiple Access
P-Prob	Primary System Model Problem
QoS	Quality of Services
RF	Radio Frequency

R.H.S	Right Hand Side
RRA	Radio Resource Allocation
SCIP	Solving Constraint Integer Programs
SINR	Signal-to-interference-noise ratio
UAV	Unmanned Aeronautical Vehicle
UMTS	Universal Mobile Telecommunications System

1

Introduction

Delivering high-capacity services over a wireless medium presents challenges, since the spectrum is limited and the demand for its access is constantly growing. For terrestrial cellular networks, the solution is to decrease the transmission range of a base station (BS) and deploy more base stations, which require backhaul interconnections. Clearly, this is a costly and difficult proposition, especially for areas with hostile geographical nature. This pressure on the radio spectrum requires moving higher in frequency to K/Ka bands (26–40 GHz), which are less heavily congested and can provide significant bandwidth. The main problem with working in K/Ka bands is that line-of-sight (LOS) or quasi-LOS propagation is needed [5].

The visibility problem can be solved using a satellite technology. This is a well-established alternative to terrestrial infrastructures that is able to serve wide areas with a cellular coverage, thus implementing frequency reuse paradigms. *Geostationary Earth Orbit* (GEOs) satellites are located at about 36,000 km away from the earth's surface. Due to the large distance from the earth's surface, GEOs have huge antenna footprints that can cover entire continents providing services to millions of users. However, being far away from the earth's surface also has major drawbacks, mainly due to the very critical free-space path loss and large propagation delays. The solution to these problems requires large antennas and sophisticated architectures and protocols at the customer receivers. Furthermore, technological constraints for on-board antennas prevent the possibility of optimizing the cell dimension on the ground, thus potentially lowering frequency reuse

1

efficiency and, consequently, overall capacity. Another type of satellites is the *Low Earth Orbit* (LEO) satellites that overcome many of the drawbacks specific to GEO satellites as they are much nearer to the earth's surface (200–1600 km). However, a single LEO satellite-based system would not be suitable for real-time transmission since the satellite is frequently out of visibility of a single ground station. In such a system, only store and forward technique could be used. If continuous coverage is required, then an entire constellation of LEO satellites would be required. Obviously this is too costly and necessitates that efficient handover schemes be used among the satellites.

A potential solution for these problems that has been considered in research so far, is the use of aerial platforms carrying communications relay payloads. These operate in a quasi-stationary position in the stratosphere layer of the atmosphere. LOS propagation paths can be provided to most users, with modest free space path loss and propagation delays, thus enabling services that take advantage of the best features of both wireless terrestrial and satellite communications. The platforms that carry these payloads were called *high-altitude platforms* (HAPs) [6].

1.1 An Overview on HAPs

HAPs are quasi-stationary aerial platforms that are meant to be located at a height of 17–22 km above the earth's surface in the stratosphere layer. Many of their pros are a combination of those in both terrestrial wireless and satellite communication systems. Some of those pros are [7]:

- Their ability to fly on demand to temporarily or permanently serve regions with unavailable telecommunications infrastructure.
- A single HAP has a large area coverage that can go up to 150 km compared to a single terrestrial cellular base station (BS) whose maximum radius, for macro cells, is in the range of 20–30 km.
- Low propagation delays compared to satellites which implies better perceived quality of service (QoS) by the users for real-time applications like voice and video.

- Stronger received signal strengths as compared to satellites and hence user terminals need not be bulky.
- Deployment time is low, since one platform and ground support are sufficient to start the service.
- Much less ground-based infrastructure compared to terrestrial cellular networks.

For the same allocated bandwidth in a specified area, terrestrial systems require a large number of base stations. On the other hand, GEO satellites have cell size limitations due to large footprints on the earth's surface. Non-geostationary satellites face handover problems and the need to deploy the entire constellation, thus requiring high launching costs to place them in orbits. In this case, HAPs seem to be an attractive choice.

HAPs can be used to serve different scenarios such as broadcast/ multicast HDTV signal, high-speed wireless access, navigation and position location systems, intelligent transportation systems, surveillance, remote sensing, traffic and environmental monitoring, emergency communications, disaster relief activities, and large-scale temporary events. In addition, one of the most promising capabilities of HAPs providing high-throughput backhaul links for ground-based pico and femto cells, thus minimizing the traffic burden in the mesh networks of the respective cellular systems. The ultimate long-term aim is to obtain integration of terrestrial, HAPs and satellite networks. A generic framework for future generation communication networks based on integration of different communication infrastructures and employing an all IP-based core network is shown in Figure 1.1.

1.2 Types of HAPs

Basically, HAPs are classified into aerostatic and aerodynamic platforms [7, 8]. This is based on the underlying physical principle that provides the lifting force. Aerostatic platforms use buoyancy to float in the air, whereas aerodynamic platforms use propulsive forces created by jet engines.

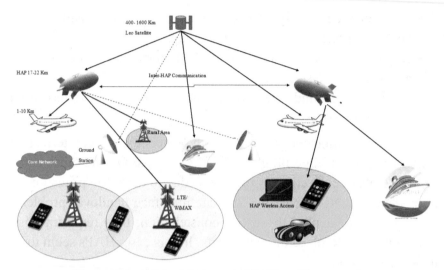

Figure 1.1   Integrated terrestrial/HAP/satellite networks.

Aerostatic platforms could be either balloons or airships. In order to provide buoyancy, they make use of a lifting gas in an envelope, most commonly hydrogen and helium. Balloons are usually unpowered platforms, and since the flight cannot be controlled easily they are usually manned. Airships, on the other hand, are normally unmanned powered platforms, capable of staying in the air for weeks and months. The main drawback of aerostatic platforms is their large size. A huge volume is needed to compensate for thin air in lower stratosphere. This causes dynamic drag during the course of a flight as well as difficulties for takeoff and landing. The large size of the platform has an advantage though, as it could accommodate larger and heavier payloads and its large area could be exploited for using solar cells to generate power.

Aerodynamic platforms rely on aerodynamic lift for flying in the air. They cannot stay in the air unless they move forward, and therefore have to be above the coverage area to keep a quasi-stationary position. Due to the density of the air at the operating altitudes of around 20 km, aerodynamic platforms require large-sized wings to obtain sufficient lift, making the radius of the circular flight of the platform in the range of a few kilometers. Due to the circular flight, the platform also requires some compensation for antenna pointing.

1.3 HAP Radio Regulations

HAP communications are subject to two main forms of regulations, radio-frequency (RF) regulations and aeronautical regulations. Since our book considers a telecommunications aspect in a HAP, we present the RF regulations only in this section. RF regulations are under the global control of the International Telecommunications Institute-Radio communications sector (ITU-R). ITU-R has allocated several frequency bands for HAPs to provide different broadband multimedia applications in millimeter-wave band and International Mobile Telecommunications (IMT)-2000 services in third generation (3G) frequency bands. ITU-R allocation specifies:

1. 300 MHz in each direction in the 47/48-GHz band world-wide on the secondary basis shared with satellites for all HAPs communications [9],
2. 300 MHz in each direction in the 31/28-GHz band also on the secondary basis in over 40 countries worldwide excluding all of Europe for fixed broadband services [10],
3. 2.1-GHz IMT-2000 band to be used for the provision of 3G services to users [11],
4. the band 5.85–7.075 GHz for HAP gateway links for fixed services [12].

In the following section, we present some of the recent published research work in the area of HAPs.

1.4 Recent Research Works in HAPs

Among the recent works in the area of HAPs, is the work done by Sudheesh et al. in [13]. In their paper, they show how spatial multi-plexing could be performed to boost the spectral efficiency. They state that in a single HAP system with multiple antennas on-board, spatial multiplexing cannot typically be achieved due to high correlation between paths. Therefore, they proposed the use of multiple spatially separated HAPs to perform precise beamforming. Due to the high altitudes and imperfect stabilization, it is challenging to acquire the necessary accurate channel state information (CSI) for precise beam-forming. To deal with this problem, the authors realize an interference alignment technique based on a multiple antenna tethered balloon that could be deployed and used as a relay between the multiple HAPs and the ground stations. In particular, a multiple-input multiple-output X network was considered in [13], and the capacity for that network was obtained in close form. The authors showed that a maximum sum-rate was obtained.

In [14], Xu et al. proposed a geometry based HAP channel model that considers the statistical and geometric properties of terrestrial environments comprehensively for the purpose of efficient deployment of HAPs. Based on their proposed channel model, they also derived the LOS transmission probability of air-to-ground communication and performed the analysis for the path loss. They also proposed an algorithm that maximizes the efficiency, in terms of the ratio of the radius of HAP footprint to inter-HAP distance. In [15], Dong et al. treated HAPs as mobile base stations and studied a placement method with guarantees on QoS and user demands in a constellation of multiple interconnected HAPs. They established QoS metrics by considering the information isolation, integrity, rate and availability. The user demand has been modeled by considering the broadband size, the population distribution density, and scale factor of the HAP network users. Based on the network coverage model, they gave out the design vector of HAPs layout optimization, i.e., number of HAPs, downlink antenna area, power of payload, longitude of HAP and latitude of HAP. Moreover, in [15] a nonlinear, nonconvex, and non-continuous combinatorial optimization model was proposed. This was solved by an improved artificial immune algorithm.

In [16], Zhang et al. considered Unmanned Aeronautical Vehicle (UAV)-enabled mobile relaying. They studied a system in which a UAV is deployed to assist in the information transmission from a ground source to a ground destination with their direct link blocked. In their paper, they study two problems, spectrum efficiency and energy efficiency maximization for that system and revealed their trade-off with the UAV's propulsion energy taken into account. The type of motion that they consider is circular motion, and the type of relaying is a decode and forward in a time-division duplex mode. They derive the optimal solutions for both problems, and show that energy efficiency maximization requires a larger circular trajectory radius than spectral efficiency maximization. Their numerical results show performance gain for mobile relaying in circular trajectory over static relaying with a fixed relay.

2

Radio Resource Allocation and User Admission Control in HAPs

Just like any wireless system, HAP needs to manage its radio resources as efficiently as possible in order to gain the maximum desired benefit. This benefit could be the system's data capacity, the number of users to be served in the system, throughput fairness among the system's users, successful packet delivery etc. One of the aspects that *radio resource allocation* (RRA) has a direct impact on, is the admission of users in the system. Simply, the availability of resources determines how many users can be admitted or served in the system. The radio resources that need to be managed for a HAP having multiple antennas using *orthogonal frequency division multiple access* (OFDMA) are:

1. the radio power,
2. the frequency subchannels,
3. the time slots over the subchannels,
4. the antennas (antenna selection).

Choosing which users to admit into the system affects the total number admitted. This is because the users have different channel conditions due to their different positions and also due to the random nature of the radio channel. For example, if a user is in a location where the received signal quality is poor, and it is to be admitted into the system, it would need too much radio power to compensate for the channel attenuation. This could lead to little remaining power that is insufficient to admit

9

other users. If that user would have not been admitted, the HAP might have been able to serve a larger number of users with good channel conditions. This is a simple example considering power only. It grows much more complex when subchannels, time slots and antenna selections are to be allocated too.

Multicasting is the transmission of the same information to a group of users instead of transmitting the same information to each user individually (unicasting). This type of transmission saves a lot of radio resources as compared to unicasting and is therefore, usually the method used to transmit same information to a group of users in any network. We can have more than one multicasting session in a HAP system, and each user may want to join more than one session at the same time. Each multicast session transmits its data on the same set of subchannels, time slots and antennas with the same power level for all users in the multicast group. RRA is needed for *admission control* (AC) of multicast sessions so that efficient admission decisions are made for users wishing to join different multicasting groups.

Since aeronautically and mechanically reliable platforms are still in the development phase, the amount of published research for telecommunication services over HAPs, particularly RRA and AC, is limited compared to other wireless systems, let alone for multicasting in specific. Moreover, most of the big HAP research projects like SHARP, Skynet, StratSat, HALO, CAPANINA, Helinet and HAPCOS [17–22] started their activities between 2000–2006, a time in which the most popular wireless interface in wireless telecommunications research was *code division multiple access* (CDMA) based *Universal Mobile Telecommunications System* (UMTS). Therefore, most of the published research in RRA and AC was for CDMA based HAPs. *Orthogonal Frequency Division Multiplexing* (OFDM) is one of the possible techniques to be used for transmission between the HAP and the users due to its well known capabilities in mitigating wireless channel impairments that result from high mobility and high transmission speeds [23]. Hence, the multiple access scheme that is expected to be used in HAPs is OFDMA. Therefore, we believe that more research in HAPs should be done considering this type of interface.

2.1 Differences between RRA in HAP Systems and Terrestrial Cellular Systems

RRA over a mutlicellular HAP system differs from conventional terrestrial cellular systems mainly due to an inherent graceful high centralization in the HAP. In the downlink, there is one common source of radio frequency (RF) power for all the cells of a given HAP, in while for a group of contiguous cells of a terrestrial cellular system, each cell has a separate BS each with independent RF power source. The same is true for the spectrum, where for the HAP the entire spectrum is shared among the HAP's cells while in conventional terrestrial cellular systems every cell uses a portion of the spectrum, depending on the frequency reuse pattern, to minimize inter-cell interference.

Also, a single HAP has the ability to have global knowledge of the channel gains of the users in all its cells at all subchannels. This is possible since all users in the HAP service area communicate with just one transmitting entity, which is the HAP. On the other hand for terrestrial cellular systems, CSI is acquired for the users in each cell by that cell's BS only. Therefore, for global CSI information to be achieved at all BSs, a transmission for each BS's CSI information over its backhaul links would be required. This is a lot of overhead signaling that would burden the network and is hence not usually performed, leading to suboptimality in multicellular RRA of terrestrial cellular systems. Furthermore, the time needed to exchange information until global CSI is achieved for a given region of a terrestrial cellular system is not guaranteed to facilitate dynamic multicellular RRA at a frame by frame basis before the CSI information at each terrestrial BS becomes obsolete.

A HAP can hence use the global CSI information it has for all users, and the fact that it has one common power and spectrum source, to centrally perform more flexible radio allocations at the HAP with full awareness of the inter-cell interference levels instantaneously on a dynamic frame by frame basis. Conventional terrestrial cellular systems would either perform RRA locally at a single cell level or if multicellular RRA is desired, a distributed approach with heavy CSI signaling overhead.

Finally, the beams of the antennas co-located on the HAP interfere with each other, as illustrated in Figure 2.1, for a single HAP system. The interference at a user in a particular cell is due to the reception of unwanted transmissions at boresight angles greater than angles that subtend the neighbor cell footprints through the mainlobes and side lobes of their antennas [24]. The collocation of the antennas allows the HAP to centrally perform electronic cell resizing by controlling the antenna beam-widths and pointing angles in an RRA problem, depending on the user distribution and/or density in a given cell, to dynamically control inter-cell interference. This is not readily possible in conventional terrestrial cellular systems.

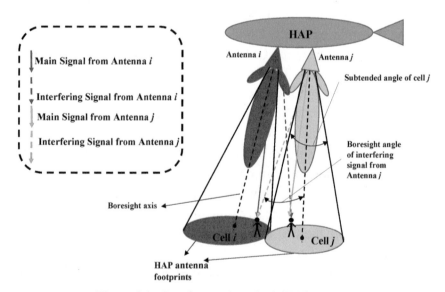

Figure 2.1   Interference in a single HAP system.

2.2 Problem Description, Description and Motivation of the Problem and the Proposed Joint AC-RRA Scheme

This book discuses a study and proposal of a novel admission control and radio resource allocation problem that was done as part of the PhD research conducted by Ahmed Ibrahim under the supervision of Prof. Alfa [1]. Mathematical formulations are derived and suitable problem-specific and structure-oriented solution methodologies are proposed. The problem discussed in this book is joint AC-RRA for an OFDMA based multiple antenna HAP, with multiple multicasting sessions of heterogenous priorities at each user in the downlink. The users have heterogeneous priorities from the service provider point of view. The QoS requirements of the admitted users and their associated multicast groups' requirements must be met, or they should not be admitted in the first place. The QoS requirements considered in this book are the signal-to-interference-noise-ratio (SINR) of a multicast session for each user and the session's minimum and maximum data rate constraints for all the multicast groups. In our earlier works in [2–4], we considered maximizing the spectrum utilization by serving the largest number of users on all the available frequency-time slots. In the extended problem in this book, we consider maximizing the number of highest priority users admissions, to their most favored sessions each.

We briefly highlight the differences between the system model we had in our earlier works [2–4] and extended one that we consider in this book. From now on, we will be referring to the system model in [2–4] as the primary problem (P-Prob), in which:

1. The concept of "cells" was adopted where each user falling within the foot print of an antenna's beam is associated with that antenna only. Hence, a user can only receive from one antenna at most and any possible antenna beam overlaps are not exploited.
2. A user can request, and hence can only receive sessions being transmitted in the cell in which the user resides.

3. All users assumed the same level of priority to the service provider, and all the sessions a given user requested were all of equal importance.

4. The spectrum utilization, i.e. the number of users each frequency-time slot can serve, was the objective to be maximized.

The extended problem (E-Prob) considers the following:

1. More flexibility by allowing transmission of a multicast session to the users in a group on more than one antenna simultaneously given an acceptable level of SINR is met for all users in the group.

2. A user can request, and hence receive, sessions being transmitted in any overlapped adjacent cell of the HAP service area, hence exploiting the possible antenna beam overlaps.

3. Each user is assumed to have heterogeneous priority levels for different multicast sessions. Also, from the service provider's point of view, the user priorities could be heterogeneous.

4. The objective is to maximize the total number of admitted users with the highest priorities, each to the sessions of the highest priority to the user.

P-Prob was the first part of our research work that was published in [2–4]. The problem was very rich and considered many different aspects that were not considered together, by other researchers in previous works for HAPs (to the best of our knowledge). Hence, we decide to go deeper in the same problem after including the extensions mentioned above to obtain further improvements. Since there could be many ways to formulate the same problem, we preferred to try to find a formulation that could be solved more efficiently than the one we obtained for P-Prob in [2–4]. We were successful in obtaining a much smaller formulation which we believe is an important achievement as any algorithm's computational effort is always proportional to the formulated problem size.

Formulating the problem using much smaller number of variables and constraints is an important step to reduce the computational effort and memory requirements by the HAP computing hardware

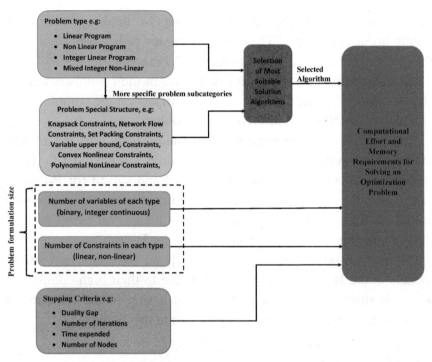

Figure 2.2   Illustration of all the factors that contribute to the computational effort and memory requirements in solving an AC-RRA optimization problem.

onboard. Figure 2.2 shows all the different aspects that contribute to the computational effort and memory requirements of solving an optimization problem for multicasting joint AC-RRA. In a general sense, the key factors of a formulation are the problem's type (e.g. linear, integer, mixed integer liner, etc.), the presence of any special structure (e.g. knapsack, transportation, quadratic, convex, etc.), and the most suitable algorithm (e.g. dynamic programming, Dijktra's algorithm, feasible directions method, branch and bound, etc.) in terms computational efficiency can be determined. Also, as shown in Figure 2.2, any algorithm's complexity is function in the problem size fed to it, and the relative numbers of different types of variables and constraints. When we have integer and continuous variables, the impact of integer variables on the computational effort is much stronger as compared to the continuous variables. The same saying goes for nonlinear constraints versus linear constraints. Therefore, since our earlier formulation for P-Prob in [2–4] has a huge number of binary variables and non-linear constraints, the significant reduction in their numbers that we achieve for E-Prob would have a crucial impact on the computational complexity encountered in solving the problem.

Since we are able to greatly reduce the problem size, we are able to extend the system model (to E-Prob) while still having a far much smaller formulation than that we obtained and solved for P-Prob in [4]. Hence, the aspect that we consider in comparing the two system models, P-Prob and E-Prob, is the formulation size for each which we illustrate in Chapter 3 to be much efficient for E-Prob.

Other than our earlier works in [2–4], we have not seen similar models in the HAP literature, probably due to their high complexity, which is the reason we decided to take a step in the direction of combining the following into one problem in this book:

1. power allocation to multicast groups,
2. subchannel allocation,
3. time scheduling,
4. multiple antenna selection,
5. user to multicast group assignments,
6. heterogeneous user priorities and
7. reusing spectrum.

The work done by Zhao et al. in [25] is the closest to our work, where a unicasting unicellular system is considered, whereas we consider a multicellular multiple-session multicasting system. Aside from power, subchannel and time allocation considered in [25], we also consider multiple antenna assignments to multicast groups and frequency reuse across the HAP cells. Finally, while the authors did not consider performing user admissions, we consider user to group admissions where each user can join more than one multicast group. Therefore, although the work done in [25] is the closest to ours, there are still many differences in the details of the system model. In the next section, an explanation is provided for how our work fits in the HAP RRA and AC, as well as the multicasting literature.

2.3 Relation between the Research Work Discussed in this Book with the Previous Works

Figure 2.3 illustrates the categorization map of the different previous works done for RRA and AC that are both relevant to the communications system we are considering, which is HAP, and the type of transmission that we are considering, which is multicast transmissions. There are three different colors used under RRA and AC in the figure. The blue color represents the research works over a CDMA-based HAP in which the majority of the HAP RRA work was done [26–31], which was mostly on unicasting. The green represents a few works [32–34] on systems similar to HAPs that use OFDMA and/or frequency division multiple accesss (FDMA), in which to the best of our knowledge, they did not consider multicasting. The red represents the work done in OFDMA multicasting [35–52], but in terrestrial cellular systems. Also, in this figure, we can see some salient shapes, which indicate the key areas that we combine together in this book.

We can see that there was no multicasting considered for OFDMA-based HAP in the literature as indicated in its corresponding shape in the green section. Recognizing this main area of research to be missing in HAP RRA and AC, motivated us to investigate it in our earlier works [2–4], extend it in this book and enhance its solution efficiency, as shown in the following chapters.

We combine three concepts from the literature:

1. central power management at the HAP level rather than the BS level from [27],
2. frequency, time and power resource allocation for OFDMA from [25] and
3. single-rate least channel gain (LCG) multicast transmissions [53].

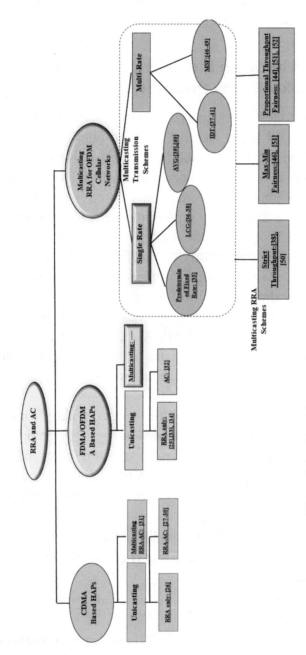

Figure 2.3   Classification of RRA and AC in previous related work.

These ideas were there individually in different works in the literature, and we combine them for our problem's system model in this book. However, each of those concepts is considered in a different context in our book. The following are different contexts for each of the individual concepts used:

1. We use the central power management at the HAP level from [27] in OFDMA-based HAP instead of a CDMA based one. We apply the concept not only to power, but to frequency and time slots.

2. Frequency, time and power resource allocation for OFDMA in [25] were considered for single-cell unicasting, but we consider the idea for multicellular multicasting in this book.

3. Single-rate LCG in the literature was obtained for the user with the least channel gain (as the technique's name suggests) in the multicast group. In this book, a similar but slightly different concept is used. Instead of using the lowest channel gain among all users in a group, we use the lowest SINR instead as it is more accurate in describing the users' quality of received signals. A user with good channel gain but too high interference would have a poor quality received signal. Moreover, in E-Prob, we define the concept of a group's highest data rate and use the highest SINR for all users in a group for a given frequency-time slot to calculate it. It is well known that variable bit-rate multicast streams, such as video, have a minimum and maximum range within which they vary. We do not want to allocate resources that yield a rate higher than the maximum possible bit-rate of the multicast session. These resources would better be used for other groups with larger number of users with poor SINR conditions.

2.4 Scope and Research Contribution in this Book

For the derived efficient formulation for E-Prob in this book, a branch and bound framework is proposed in which we use linear outer approximation by McCormick underestimators as a relaxation for the formulated mixed binary quadratically constrained program [54] and *mixed integer linear programming* techniques. Different branching schemes for the branch and bound scheme are considered and their performances are evaluated by numerical experiments [55]. Also, a reformulation technique that linearizes a certain type of quadratic constraints in the formulation is used and computational experiments are conducted to evaluate the performance with and without the linearization scheme. Domain propagation methods, separating cuts and heuristics are also used in solving the formulate problem of E-Prob [56]. The parameters used for performance comparison in computational experiments are:

1. the duality gap,
2. the number of branch and bound (BnB) nodes needed,
3. the number of iterations needed,
4. the average number of iterations per BnB node,
5. the number of instances for which a feasible solution is found,
6. the time needed to find the first feasible solution, and
7. the value of the objective function.

The reasons for using these criteria for the performance evaluation is as follows. An important aspect in solving an optimization problem for AC-RRA over HAP is to find the best possible solution quickly before the time variant radio fading channel changes, which would lead to a change in the optimization problem's constraint coefficients. Furthermore, due to the HAP's limited hardware resources onboard, it is desired to keep the memory usage as low as possible. The speed of obtaining the solution is measured by either the number of iterations, the *central processing unit* (CPU) time consumed or both. The memory usage is directly dependent on the size of the branch and bound tree. Therefore, the smaller the number of nodes, the smaller the memory needed. The quality of the solution is measured using the duality gap, since according to weak duality, the lower and upper bounds approach each other as better quality solutions are obtained in the tree, whose objective value is given by the value of the lower bound.

The branch and bound algorithm is a generic framework that is mostly used for problems that involve binary or integer variables. It is a divide and conquer scheme that divides the problem into smaller simpler problems, known as nodes. The *root* node is the whole problem before division, while the rest of the nodes are smaller subproblems that have either been solved or still need to be solved. A *leaf* node is a subproblem that either has not yet been processed or one that has been processed and *pruned*. Any pruned node has no descendants. The *bounding* step avoids complete enumeration of potential solutions of the problem by pruning nodes. A node gets pruned when it proves that none of its descendants can give a better primal feasible solution than the one found so far during the course of the algorithm. The details of the procedure are provided in Chapter 4. The main benefit of using branch and bound-based solution methods is that if we decide to stop the algorithm before obtaining the optimum solution due to time limitations, we are left with the best solution found. This is a sub-optimal solution, whose goodness is determined by the values of the upper and lower bounds. Also, it is possible for the branch and bound algorithm to terminate once it obtains an acceptable sub-optimal solution that is within any desired % from the optimal. Hence, we can say that the branch and bound-based approach can be used to return sub-optimal solutions within an acceptable nearness to the optimal or, the best found solution in a maximum tolerable solution time.

The main purpose of separating cuts is to tighten the relaxation of a problem and hence produce sharper bounds that help in early pruning. The branching procedure is the selection of the variable to branch on to create the new descendant nodes. A branching scheme that yields best dual bounds is desirable as it leads to early pruning in the tree. In the research problem under consideration, techniques from [56] and [57] are used, where an outer approximation is generated by linear underestimation of the non-convex quadratic constraints to relax the problem's feasible region. The problem becomes a *mixed binary linear program* (MBLP), and hence linear programming (LP) techniques can be used within a branch-and-cut algorithm. The BnB algorithm recursively splits the problem into smaller subproblems, thereby creating a branch-ing tree. Different branching schemes are used for that purpose. At each node, domain propagation is performed to exclude further values from the variables' domains, and a relaxation may be solved to achieve an upper (dual) bound. The relaxation is then strengthened by adding further valid constraints, which cut off the optimum of the relaxation. Primal heuristics are integrated in the BnB procedure to improve the lower (primal) bound. These are explained in details in Chapter 5.

Since the problem considered is quite different from the other problems in the HAP literature, and even the terrestrial wireless mul-ticasting in OFDMA systems, we believe the comparison of our work with those would not provide us with sufficiently useful conclusions. Basically, the decisions to be made, the QoS constraints and even the objective function are all different. The comparisons that we per-form, however, are in the sizes of the formulation derived here versus the one in the earlier works [2–4] and the performance of different solution components like branching rules, the effect of introducing domain propagations, separating cuts and heuristics, etc. Comparison of different algorithmic components allows us to conclude the most efficient choices to use for solving the new efficient formulation whose derivation is presented in this book. We compare the two formulations in Chapter 3 and illustrate that the one derived in this book is more efficient.

For the different choices for each of the solution algorithm's components, we experimentally perform the comparisons to find the most efficient choices in terms of solution goodness, speed (number of iterations), etc. Three different experiment sets are provided. The first experiment set compares the performance of activating-versus-deactivating a reformulation linearizion technique for a certain type of quadratic constraints, at the presolving phase. The second experiment set compares the performance of different branching techniques discussed in Chapter 4. Accordingly, at the end of Chapter 4, we conclude with the most suitable choices for branching and presolving reformulation linearizion, which we suggest. The third experiment set is given in Chapter 5 and compares the performance of different combinations of domain propagation, cutting planes and heuristics.

3

Multicasting in a Single HAP System: System Model and Mathematical Formulation

In this chapter, the system model for the problem considered in this book is presented in Section 3.1. The system model in our earlier published works [2–4] is extended and an elaboration on the differences between the previous system model and the extended system model is made. The key differences between the equations that describe E-Prob and P-Prob are given in Section 3.2 and then the formulation of E-Prob is provided and discussed in Section 3.3. We show how the formulation can be transformed to an equivalent mixed binary polynomial constrained program in Section 3.4 and then how that is reduced to an equivalent mixed binary quadratically constrained program in Section 3.5. For that resulting formulation, we apply the solution procedure explained in Chapters 4 and 5. Finally, a comparison between the sizes of the derived E-Prob formulation and that of P-Prob is given in Section 3.6 of this chapter.

3.1 System Model

In this section, the extended system model (E-Prob), for multicasting AC-RRA in an OFDMA HAP system is provided. A simple standalone HAP architecture [7] is considered. A user can request to receive, and be admitted to, sessions which are not only being transmitted within the cell it resides in, but also those being transmitted in neighboring cells, if the SINR is acceptable. This means that after the admission is done, a user can belong to multicast groups in different cells simultaneously.

The main difference between E-Prob and P-Prob is that we no longer adopt the concept of user association to "cells" as in terrestrial cellular systems. Instead, a multicast group could actually receive transmission on more than one antenna on different frequency-time slots simultaneously. P-Prob did not allow that since it adopted the concept of cells where a user can receive only from the antenna that illuminates the cell in which it resides in. In P-Prob, a group of users that receive the same multicast session in different cells were considered to be separate groups while in E-Prob, all users receiving the same multicast session are considered in the same group regardless of the antennas they are receiving on. The second difference is that P-Prob considered that a user can only receive multicast sessions being transmitted in the cell it resides. If a user would like to receive a session that is being transmitted in another cell but is not currently being transmitted in the cell it belongs to, then the user would not be able to receive the session. In E-Prob, however, a user can receive a multicast session being transmitted in a neighboring cell, if it is not being transmitted in the cell in which the user resides in. This is possible if the two transmissions on antennas i and j are performed on separate sets of frequency-time slots. The possibility increases for users near the cell boundaries, especially because antennas footprints do not have deterministic contours outside which the received power is zero and hence the received powers from each could be overlapping in certain areas as shown in Figure 3.1. Finally, E-Prob considers different multicast session priorities for user-to session admissions where each user could have different priority levels from the service provider's perspective, and each session has different priority degrees for different users. We aim at maximizing the number of highest priority user-to-session admissions, instead of giving all the users homogeneous priority levels as in P-Prob.

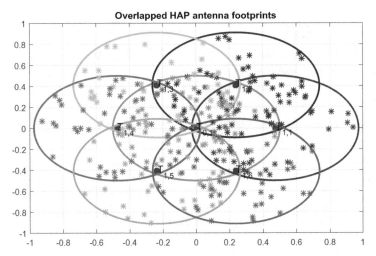

Figure 3.1 Illustration of the HAP antenna beam overlaps.

The set of users that get admitted to receive a multicast session m are considered a multicast group with the same index of the session, m. The HAP has multiple antennas over which the multicast streams are transmitted to the service area. A user can request to receive more than one session and hence, may be admitted to (allowed to receive) one or more of the requested sessions. This means that after the admission is done, a user can belong to more than one multicast group. Any two multicast sessions may not be transmitted on the same resource trio combination (i, c, t) to avoid inseparable signal interference, where i is the antenna index, c is the subchannel index and t is the time slot index. For a frequency-time slot (c, t) to be assigned to a particular user to receive session m on antenna i, it has to satisfy a minimum SINR threshold $\gamma_{m,i}^{th}$ to guarantee an acceptable bit-error-rate performance. $\gamma_{m,i}^{th}$ could be different across the sessions and antennas depending on the possibly different modulation and channel coding schemes. The main notations used for the problem formulation in this chapter are provided in Table 3.1.

Table 3.1   Notation Definitions for E-Prob Formulation

Notation	Definition
M	is the number of multicast sessions in the HAP service area.
S	is the number of HAP antennas onboard.
K	number of users in the service area.
C	is the number of available subchannels.
T	total number of time slots available in the OFDMA frame.
ΔB	is the subchannel bandwidth.
ΔT	is one time slot duration.
F	is the OFDMA frame duration.
σ^2	is the additive white Gaussian noise power per subchannel.
$p_{m,i,c,t}$	is the value of the HAP power assigned for multicast session m on antenna i in the frequency-time slot (c,t).
$g_{i,k,c,t}$	channel gain between antenna i and user k on frequency-time slot (c,t).
$\lambda_{m,k}$	is a binary constant that indicates whether user k has requested to join session m.
$\phi_{m,k}$	is a binary variable that indicates whether a user k gets assigned to receive multicast session m.
$\rho_{m,k}$	is a positive integer constant that represents the priority level for admitting user k to the group (i.e. session) m.
θ_m	is a binary variable that indicates whether session m receives any resources, or equivalently, whether any user gets assigned to receive the session's transmission.
$y_{m,i,c,t}$	is a binary variable that indicates whether the trio combination (i,c,t) is assigned for session m.
$\hat{M}$	is a very large arbitrary number.
$\gamma_{m,i}^{th}$	is the SINR value that satisfies a desired target bit-error-rate (BER) for session m on antenna i. Different sessions transmitted on different antennas may be modulated and coded differently thus requiring different SINR thresholds.

Figure 3.2 shows the power $p_{m,i,c,t}$ for session m being assigned to the trio (i,c,t). The antenna, frequency and time resources are represented graphically by three dimensions where the antenna dimension is not necessarily orthogonal to the frequency-time plane due to the possibility of antenna foot print overlaps. Orthogonality here means the absence of interference between any pair of trios (i,c,t) represented by the small cubes in the figure. HAP power is allocated to each of

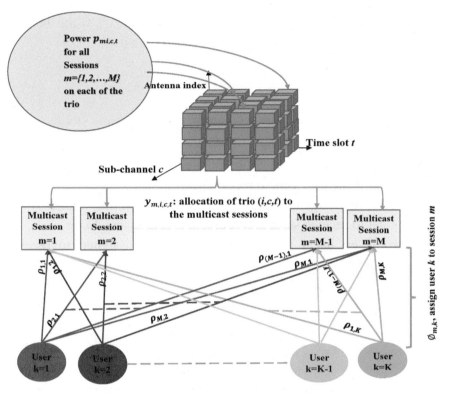

Figure 3.2 Illustration of the multicasting AC-RRA in E-Prob.

the trio cubes for the different multicast sessions being transmitted to the service area. The "cubes" are assigned to the different multicast groups and the users in the HAP service area are assigned to these groups according to their priority value $\rho_{m,k}$, *quality-of-service* (QoS) requirements and availability of resources.

For E-Prob, there are two definitions associated with a group's data rate. The minimum rate of the group is defined as:

$$\hat{R}_m^{min} = \sum_{i=1}^{S} \sum_{c=1}^{C} \sum_{t=1}^{T} r_{m,i,c,t}^{min}, \qquad (3.1)$$

where $r^{min}_{m,i,c,t}$ is the rate of session m over the trio (i, c, t) for the user with the minimum SINR on (i, c, t) and is given as:

$$r^{min}_{m,i,c,t} = \frac{\Delta B \Delta T}{F} log_2 \left(1 + \min_k x_{m,i,k,c,t} \right),$$ (3.2)

where ΔB is the subchannel bandwidth, ΔT is the time slot duration, F is the OFDMA frame length duration and $x_{m,i,k,c,t}$ either:

- takes the value of the SINR of user k on the trio combination (i, c, t) if the user gets to receive session m,
- takes a very large number $\hat{M}$ (theoretically infinity) if user k does not get to receive session m but some other users do, or
- zero if no users in the service area are assigned to receive session m.

hence $x_{m,i,k,c,t}$ can be expressed as:

$$x_{m,i,k,c,t} = \frac{P_{m,i,c,t} \left[g_{i,k,c,t} + (1 - \phi_{m,k}) \hat{M} \right]}{\sum_{m=1}^{M} \sum_{\forall i' \neq i} g_{i',k,c,t} P_{m,i',c,t} + \sigma^2}.$$ (3.3)

where $g_{i,k,c,t}$ is the channel gain on each antenna-frequency-time trio combination (i, c, t) for user k and $\phi_{m,k}$ is a binary variable indicating user-to-session admission for user k.

The channel gains $g_{i,k,c,t}$ depend upon the instantaneous values of large-scale fading and small-scale fading. In a HAP system, large-scale fading is a result of free space path loss and attenuation due to gas absorption and water vapor [58]. Small-scale fading is acceptably modeled as Ricean fading due to the presence of line of sight rays from the HAP to most of the locations in the HAP service area [5]. The channel gain $g_{i,k,c,t}$ between antenna i and user k on the frequency-time slot (c, t) can hence be given as:

$$g_{i,k,c,t} = \left(\frac{\check{C}_{light}}{4\pi d_k f_c} \right)^2 .G_H (\varpi_{i,k}) .G_k^u . \frac{1}{A(d_k)} .\varphi_{k,c,t}$$ (3.4)

where

- $G_H(\varpi_{i,k})$ is the gain seen at an angle $\varpi_{i,k}$ between user terminal k and antenna i boresight axis.
- d_k is the distance between the HAP and user terminal k, $\check{C}_{light}$ is the speed of light and f_c is the carrier frequency.
- $A(d_k)$ is the attenuation due gas absorption, clouds and rain. This depends on the distance between the HAP and each user k in the service area.
- G_k^u the antenna's gain of user terminal k.
- $\varphi_{k,c,t}$ is the Ricean small scale gain in frequency-time slot (c,t) for user terminal k.

We also define the maximum rate of a multicast group as:

$$\hat{R}_m^{max} = \sum_{i=1}^{S}\sum_{c=1}^{C}\sum_{t=1}^{T} r_{m,i,c,t}^{max}, \tag{3.5}$$

where $r_{m,i,c,t}^{max}$ is the data rate of session m over the trio combination (i,c,t) which is defined to be the data rate of the user with maximum SINR on (i,c,t) and is given as:

$$r_{m,i,c,t}^{max} = \frac{\Delta B \Delta T}{F} log_2\left(1 + \max_k t_{m,i,k,c,t}\right), \tag{3.6}$$

where $t_{m,i,k,c,t}$ either:

- takes the value of the SINR of user k on the trio combination (i,c,t) if the user gets to receive session m, or
- is zero if user k does not get to receive session m.

hence $t_{m,i,k,c,t}$ can be expressed as:

$$t_{m,i,k,c,t} = \frac{g_{i,k,c,t}P_{m,i,c,t}\phi_{m,k}}{\sum_{m=1}^{M}\sum_{\forall i' \neq i} g_{i',k,c,t}P_{m,i',c,t} + \sigma^2}. \tag{3.7}$$

3.2 Key Differences in the Fundamental Equations that Describe E-Prob and P-Prob

In our earlier work in [4], since the spacial dimension (i.e. antenna selection) was not considered, the data rate for a multicast group m

was defined as:

$$\hat{R}_m = \sum_{c=1}^{C} \sum_{t=1}^{T} r_{m,i,c,t}^{min} \tag{3.8}$$

which did not sum the data rates on the different antennas as Equations (3.1) and (3.5) do for E-Prob. This ruled out the possible advantage that users in a group in cell i can receive a session being transmitted on one antenna illuminating a neighboring cell, but not being transmitted in the one it resides. It also prohibited a multicast group of users from exploiting the inherent spacial diversity provided by the multiple collocated onboard antennas, where a resource unit is the trio antenna-frequency-time (i, c, t) allowing a group to receive from more than one antenna simultaneously provided the SINR is above an acceptable threshold for all the group's users. In E-Prob, even if the group of users was to receive a session on only one antenna, the system has the flexibility to select which antenna to receive on, as long as more than one antenna stream the session. This was not permitted by the formulation (O) of P-Prob in [2–4], that was based on Equation (3.8). Constraint set C2 of formulation (O) in [4] is given by:

$$z_{m,i,k,c,t,} = z_{m,i,k,c',t'}, \quad \forall c,t : x_{m,i,c,t} = 1,$$
$$\forall c',t' : x_{m,i,c',t'} = 1, \forall m,i,k$$

where:

- $z_{m,i,k,c,t}$ is the set of binary decision variables that indicated whether a user k got admitted to receive transmission m from the antenna covering cell i on a frequency-time slot (c, t),
- $x_{m,i,c,t}$ was a binary decision variable that indicated whether a frequency-time slot (c, t) got assigned for the group $N_{m,i}$ and
- $N_{m,i}$ is the group of multicast users residing in cell i, receiving session m and can only receive transmission from antenna i.

This constraint ensured that the users assigned to receive session m in cell i over a set of frequency-time slots should be the same in each of those assigned slots, since in multicasting, the same set of resources are shared by the set of users in the multicast group. The constraint at the same time enforces the set of users receiving session m in cell i, to receive it from the antenna of that cell only, and treated groups receiving the same session in another cell i' as a different group $N_{m,i'}$. Another constraint set in formulation (O) in [4] that did not take into account antenna selection and possible reception from more than one antenna simultaneously, is constraint set C1 given by:

$$z_{m,i,k,c,t} = \begin{cases} \{0,1\} & if\ \lambda_{m,i,k} = 1, \\ 0 & \text{otherwise}, \end{cases} \quad \forall m, i, k, c, t$$

where $\lambda_{m,i,k}$ was a binary constant that indicated whether user k resided in cell i and sent a request to receive session m. Since P-Prob considered no cell overlaps, a user could only physically reside in one cell and hence $\lambda_{m,i,k}$ is equal to 1 for exactly one i. This also prevented the user from receiving transmission from any other antenna except the one for which $\lambda_{m,i,k} = 1$, if the user got admitted.

Note that Equation (3.8) was used to impose upper and lower data rate constraints on session m in cell i as

$$R_m^{min} \leq \sum_{c=1}^{C} \sum_{t=1}^{T} r_{m,i,c,t}^{min} \leq R_m^{max} \quad \forall k : z_{m,i,k,c,t} \neq 0, \forall m, i \quad (3.9)$$

which used the data rate $r_{m,i,c,t}^{min}$ to define the data rate of session m in cell i over the frequency-time slot (c, t) as that of the user with the poorest SINR. For the lower data rate constraint, this guarantees that all users in the group receive a data rate greater than the minimum. The definition of a multicast group data rate in Equation (3.8) was also used to enforce a maximum rate R_m^{max} constraint. However, it was noticed that the upper data rate constraint may not be necessarily satisfied for all users in a multicast group on a particular frequency-time slot if we use the data rate of the user with the poorest SINR in the group to solely describe the group's data rate. This was the reason we introduced $r_{m,i,c,t}^{max}$ and $\hat{R}_m^{max}$ in Equations (3.6) and (3.5).

In P-Prob, the objective function was given by [4]:

$$\max \sum_{m=1}^{M} \sum_{i=1}^{S} \sum_{k=1}^{K} \sum_{c=1}^{C} \sum_{t=1}^{T} z_{m,i,k,c,t}, \quad (3.10)$$

which captured the sum of the users for every multicast group $N_{m,i}$ served by each frequency-time slot (c, t) which we defined to be the spectral utilization. The objective function for P-Prob does not consider user-session priorities. The next section introduces the objective function of E-Prob and explains it to show the differences with that of P-Prob.

3.3 Formulation of E-Prob

This section illustrates an efficient formulation for the extended problem. We achieve a more efficient formulation than we would have, had we just directly extended our earlier formulation in [2–4]. The number of variables and functional constraints in the new formulation are greatly reduced, which we believe to be an important achievement, especially that this was achieved for an extended model. Using the newly defined variables $\phi_{m,k}$, θ_m and $y_{m,i,c,t}$, the E-Prob problem's formulation takes into account:

- the same QoS, resource and multicast transmission requirements as in the P-Prob,
- as well as the differences in the extended system model explained earlier in Section 3.1.

The key thing that enabled us to obtain a smaller formulation, is replacing the variable $z_{m,i,k,c,t}$ in formulation (OP1) in [2–4] with the two variables $\phi_{m,k}$ and $y_{m,i,c,t}$. The formulation is given below, and an interpretation for each constraint set is provided right after:

$$\max_{\phi_{m,k},\theta_m,y_{m,i,c,t},p_{m,i,c,t}} \sum_{m=1}^{M} \sum_{k=1}^{K} \rho_{m,k}\phi_{m,k} \qquad (\mathcal{HAP}^{Eff})$$

s.t.

$$C1 : \phi_{m,k} \leq \lambda_{m,k}, \quad \forall m, k$$

$$C2 : \sum_{m=1}^{M} y_{m,i,c,t} \leq 1, \quad \forall i, c, t$$

$$C3 : \sum_{i=1}^{S} \sum_{c=1}^{C} \sum_{t=1}^{T} y_{m,i,c,t} \geq \phi_{m,k}, \quad \forall m, k$$

$$C4 : y_{m,i,c,t} \leq \sum_{k=1}^{K} \phi_{m,k}, \quad \forall m, i, c, t$$

$$C5 : P_{PF}^{Total} y_{m,i,c,t} \geq p_{m,i,c,t}, \quad \forall m, i, c, t$$

$$C6 : \sum_{m=1}^{M} \sum_{i=1}^{S} \sum_{c=1}^{C} p_{m,i,c,t} \leq P_{PF}^{Total}, \quad \forall t$$

$$C7 : p_{m,i,c,t} \geq 0, \quad \forall m, i, c, t$$

$$C8 : \frac{g_{i,k,c,t} p_{m,i,c,t} + (1 - \phi_{m,k}) \hat{M}}{\sum_{m=1}^{M} \sum_{i' \neq i} g_{i',k,c,t} p_{m,i',c,t} + \sigma^2} \geq y_{m,i,c,t} \gamma_{m,i}^{th}, \quad \forall m, i, k, c, t$$

$$C9 : \sum_{i=1}^{S} \sum_{c=1}^{C} \sum_{t=1}^{T} y_{m,i,c,t} \leq SCT.\theta_m, \quad \forall m$$

$$C10 : \sum_{i=1}^{S} \sum_{c=1}^{C} \sum_{t=1}^{T} y_{m,i,c,t} \geq \theta_m, \quad \forall m$$

$$C11 : \theta_m R_m^{min} \leq \sum_{c=1}^{C} \sum_{t=1}^{T} \sum_{s=1}^{S} \frac{\Delta B \Delta T}{F} log_2 \left(1 + \min_{k} x_{m,i,k,c,t} \right), \quad \forall m$$

$$C12 : \sum_{c=1}^{C} \sum_{t=1}^{T} \sum_{s=1}^{S} \frac{\Delta B \Delta T}{F} log_2 \left(1 + \max_{k} t_{m,i,k,c,t} \right) \leq R_m^{max}, \quad \forall m$$

The interpretation of the objective function and constraints in $\mathcal{HAP}^{Eff}$ is as follows:

- The objective function represents a weighted sum of all admissions of different users over all sessions. The larger weights force user-to-session admissions of highest priorities as long as the QoS, SINR and group data rate requirements can be satisfied. This is different from the objective function in [2–4] which sums all the users, assuming homogeneous priorities, in all the frequency-time slots across all HAP cells.

- $C1$ ensures that if user k does not request to receive session m (i.e. $\lambda_{m,k} = 0$), then the user can never be assigned to receive it (i.e. $\phi_{m,k}$ is set to zero). This constraint set is somehow similar to constraint set $D1$ of formulation (OP1) in [2–4] (P-Prob), yet consists of $M.K$ constraints versus $M.S.K.C.T$ in $D1$. The functional difference is that $D1$ of P-Prob ensures that the user can be admitted to receive session m when:
  - user k is in cell i, and
  - session m is being transmitted in cell i.

  In E-Prob, we do not have these two restrictions.
- $C2$ ensures that a given trio combination (i, c, t) can at most be assigned to one multicast group (session). This is equivalent to constraint set $D5$ of formulation (OP1) in [2–4], yet consists of a much smaller number of constraints as will be shown later in Section 3.6.
- $C3$ ensures that user k can be assigned to multicast group m only when the session gets assigned at least one resource trio combination (i, c, t). Besides C4, this constraint set is also required in $\mathcal{HAP}^{Eff}$ to connect the two sets of variables $\phi_{m,k}$ and $y_{m,i,c,t}$. These were not required in formulation (OP1) in [2–4] since $\phi_{m,k}$ and $y_{m,i,c,t}$ were captured in a single variable, $z_{m,i,k,c,t}$.
- $C4$ ensures that if no users are assigned to session m, then no resource trios (i, c, t) should be allocated to the group.
- $C5$ ensures that if the trio combination (i, c, t) is not assigned for session m, then the power level assigned for group m on (i, c, t) should be forced to zero. This is equivalent to constraint set $D10$ in formulation (OP1) in [2–4]. However, each constraint in $C5$ of

$\mathcal{HAP}^{Eff}$ has only two variables compared to $K + 1$ variables in each constraint of $D10$ in formulation (OP1) in [2–4].

- $C6$ ensures that the total power at a given time slot assigned for all multicast groups on all antenna-frequency (i, c) pairs must be limited to the total available HAP power. This is exactly the same constraint as $D9$ in formulation (OP1) in [4].

- $C7$ ensures that the power values $p_{m,i,c,t}$ are all non-negative, which is exactly the same as $D11$ in formulation (OP1) in [2–4].

- $C8$ is a constraint set that enforces the SINR of user k receiving session m to be greater than a threshold value $\gamma_{m,i}^{th}$, in order to admit the user to group m. There are three possibilities for each of the constraints in the set, which are explained as follows:

  1. If the trio combination (i, c, t) is not assigned to session m (i.e. $y_{m,i,c,t} = 0$), constraint $C5$ forces the power variable $p_{m,i,c,t}$ to be zero. This makes the left hand side (L.H.S) in constraint ($C8$) either equal to the very large number $\hat{M}$, or equal to zero, depending on the value of $\phi_{m,k}$. Both cases satisfy the inequality making the constraint redundant.

  2. If the trio (i, c, t) is assigned to session m (i.e. $y_{m,i,c,t} = 1$), but user k is not assigned to receive m (i.e. $\phi_{m,k} = 0$), the power variable $p_{m,i,c,t}$ could take any non-zero value. In this case, the term in the numerator of the L.H.S becomes greater than or equal to the very large number $\hat{M}$ making the constraint redundant.

  3. For $y_{m,i,c,t} = 1$, if user k is to get admitted for session m, then $\phi_{m,k} = 1$. In this case, the term on the L.H.S of the constraint equivalent to the SINR for session m to user k since the numerator becomes the product of the power variable $p_{m,i,c,t}$ times the channel gain of the user on the trio combination (i, c, t). The right hand side (R.H.S.) also becomes equal to the acceptable threshold value, $\gamma_{m,i}^{th}$, for session m on antenna i. In this case the SINR constraint over the trio combination (i, c, t) comes into effect for user k and session m.

Constraint set $C8$ in $\mathcal{HAP}^{Eff}$ is functionally equivalent to $D3$ in formulation (OP1) in [2–4].

- $C9$ and $C10$ together ensure that only if there are any resources being assigned for session m, then this must set the variable $\theta_m = 1$, otherwise $\theta_m = 0$ is enforced. This is needed for the minimum data rate constraint $C12$. Constraint sets $C9$ and $C10$ have no equivalent constraint sets in formulation (OP1) in [2–4].

- $C11$ ensures the minimum rate R_m^{min} of a multicast session is satisfied. We use the definition of the minimum rate of a multicast group given in Equation (3.1). There are four possibilities for $x_{m,i,k,c,t}$ (defined by Equation 3.3), which are explained as follows:

  1. $y_{m,i,c,t} = 0$ *and* $\phi_{m,k} = 0$. In this case, constraints $C5$ will force the power variable $p_{m,i,c,t}$ to be zero which results in, $x_{m,i,k,c,t} = 0$ and $\min_{k} x_{m,i,k,c,t} = 0$ giving a rate of zero on the trio combination (i, c, t).

  2. $y_{m,i,c,t} = 0$ *and* $\phi_{m,k} = 1$. This would have exactly the same result as the first case, a rate of zero on that trio combination (i, c, t) for the same reasons.

  3. $y_{m,i,c,t} = 1$ *and* $\phi_{m,k} = 0$. In this case $x_{m,i,k,c,t} = \infty$ theoretically, which ensures that for that particular user, its SINR value is never returned by the term $\min_{k} x_{m,i,k,c,t}$. There are definitely other users who have $\phi_{m,k} = 1$, according to constraint $C4$, from which the least SINR on (i, c, t) is returned by $\min_{k} x_{m,i,k,c,t}$.

  4. $y_{m,i,c,t} = 1$ *and* $\phi_{m,k} = 1$, in this case $x_{m,i,k,c,t} = \dfrac{p_{m,i,c,t} g_{i,k,c,t}}{\sum_{m=1}^{M} \sum_{\forall i' \neq i} g_{i',k,c,t} p_{m,i',c,t} + \sigma^2}$ which is the SINR of session m experienced by user k over the trio combination (i, c, t). Therefore, $\min_{k} x_{m,i,k,c,t}$ would return the minimum SINRs among all users in group m over (i, c, t).

The variable θ_m ensures that the constraint is not in effect in the case that no resources are allocated at all for session m, i.e. $\theta_m = 0$. This constraint set extends the lower bound constraint set for $C4$ in formulation (O) in [4] by summing the data rate of session m over all the HAP antennas. It is worth noting that for P-Prob, constraint set $D2$ in the reformulation (OP1) in [4] enforces all users to receive multicast sessions from only one antenna, which is the antenna that covers the cell they reside in.

- $C12$ ensures that the maximum rate requirement of the group (or session) m, defined by Equation (3.5), is satisfied. The possibilities for $t_{m,i,k,c,t}$, defined by Equation (3.7), are explained as follows:

  1. For the case $y_{m,i,c,t} = 0$, no matter what the value of $\phi_{m,k}$ is, the power variable $p_{m,i,c,t}$ is forced to zero by constraint $C5$, therefore we get $t_{m,i,k,c,t} = 0 \; \forall \; k$, and $\max_{k} t_{m,i,k,c,t} = 0$.

  2. For the case $y_{m,i,c,t} = 1$, and user k is not assigned to group m, i.e. $\phi_{m,k} = 0$. In this case, $t_{m,i,k,c,t}$ returns zero but the term $\max_{k} t_{m,i,k,c,t}$ returns the highest SINR, over (i, c, t), amongst all users assigned to session/group m. We are sure that if $y_{m,i,c,t} = 1$ then there is at least one user who has $\phi_{m,k} = 1$ according to constraint set $C5$.

  3. For the case $y_{m,i,c,t} = 1$ and user k assigned to the group m, i.e. $\phi_{m,k} = 1$, $t_{m,i,k,c,t} = \dfrac{p_{m,i,c,t}g_{i,k,c,t}}{\sum_{m=1}^{M} \sum_{\forall i' \neq i} g_{i',k,c,t} p_{m,i',c,t} + \sigma^2}$ and the term $\max_{k} t_{m,i,k,c,t}$ returns the highest SINR over (i, c, t) amongst all users assigned to session/group m.

Constraint set $C12$ in $\mathcal{HAP}^{Eff}$ is different from its equivalent upper bound data rate constraint set $C4$ in formulation (O) in [2–4] in two aspects. The first aspect is that it utilizes the newly introduced concept of maximum multicast group data rate mentioned earlier in this section and given by Equations 3.5 and 3.6. In this way, it is guaranteed that no user in any multicast group can have a data rate greater than R_m^{max}. Constraint set $C4$ in formulation (O) in [4] on the other hand uses the data rate of the user with the poorest channel conditions to define the group's data capacity, and it is that data rate that is enforced to be no more than R_m^{max}. This could lead to users with good SINR conditions in a group receiving a rate greater than R_m^{max}, which constraint set $C12$ in $\mathcal{HAP}^{Eff}$ makes sure does not happen. The second difference is that since E-Prob allows the users in a group m to receive the multicast transmission on more than one antenna simultaneously, then the maximum data rate of the group is obtained by summing all the group's data rates over all the antennas. This was not considered in formulation (O) in [2–4].

It is worth mentioning that the SINR constraint set $C8$ in $\mathcal{HAP}^{Eff}$ ensures that for a given multicast session m, no more than one antenna can be used to transmit the session over the same frequency-time slot (c, t). This is possible since in the L.H.S. of the constraint set, the interference terms in the denominator include received copies of the same desired session m from the other antennas of the HAP from which the user is not meant to receive in the frequency-time slot (c, t). The entire constraint set $C8$ guarantees that if the SINR requirement is satisfied by receiving a session on one antenna in slot (c, t), then this could not be possible simultaneously over any other antenna for slot (c, t) given the assumption $\gamma_{m,i}^{th} \geq 1$.

As we can see, the problem formulation labeled $\mathcal{HAP}^{Eff}$ is a binary mixed nonlinear constrained problem. Constraint set $C8$ has a special structure of being a mixed binary quadratic constraint set that consists only of bilinear terms. Constraint sets $C11$ and $C12$ are non linear mixed binary constraints with *min* and *max* terms respectively that complicate them further. In Section 3.4, the reformulation techniques are used to eliminate the *min-max* terms and replace those constraints with multivariate polynomial constraints. Then we show how the polynomial constraints are reduced to multivariate quadratic constraints that consist only of bilinear terms in Section 3.5.

3.4 Reducing the Formulation to a Mixed Binary Polynomial Constrained Problem

In this section, we show how the constraint sets $C11$ and $C12$ in $\mathcal{HAP}^{Eff}$ are replaced by mixed binary polynomial constraints (*MBPCs*), some of which are quadratic. For constraint set $C11$ in $\mathcal{HAP}^{Eff}$, the constraint can be rewritten in the form:

$$\log_2\left[\prod_{i=1}^{S}\prod_{c=1}^{C}\prod_{t=1}^{T}\left(1+\min_{k}\frac{p_{m,i,c,t}[g_{i,k,c,t}+(1-\phi_{m,k})\hat{M}]}{\sum_{m=1}^{M}\sum_{\forall i'\neq i}g_{i',k,c,t}p_{m,i',c,t}+\sigma^2}\right)\right]\geq\frac{\theta_m R_m^{min}F}{\Delta B\Delta T},\ \forall m$$
$$(3.11)$$

Taking exponential of 2 for both sides of the constraint, we get:

$$\prod_{i=1}^{S}\prod_{c=1}^{C}\prod_{t=1}^{T}\left(1+\min_{k}\frac{p_{m,i,c,t}[g_{i,k,c,t}+(1-\phi_{m,k})\hat{M}]}{\sum_{m=1}^{M}\sum_{\forall i'\neq i}g_{i',k,c,t}p_{m,i',c,t}+\sigma^2}\right)\geq 2^{\frac{\theta_m R_m^{min}F}{\Delta B\Delta T}},\ \forall m.$$
$$(3.12)$$

The right hand side of the constraint can be rewritten to give the constraint:

$$\prod_{i=1}^{S}\prod_{c=1}^{C}\prod_{t=1}^{T}\left(1+\min_{k}\frac{p_{m,i,c,t}[g_{i,k,c,t}+(1-\phi_{m,k})\hat{M}]}{\sum_{m=1}^{M}\sum_{\forall i'\neq i}g_{i',k,c,t}p_{m,i',c,t}+\sigma^2}\right)\geq\hat{R}_m^{min}\theta_m+(1-\theta_m),\ \forall m,$$
$$(3.13)$$

where $\hat{R}_m^{min}=2^{\frac{R_m^{min}F}{\Delta B\Delta T}}$. Then we introduce the auxiliary variables $w_{m,i,c,t}$ for the terms

$$\left(1+\min_{k}\frac{p_{m,i,c,t}\left[g_{i,k,c,t}+(1-\phi_{m,k})\hat{M}\right]}{\sum_{m=1}^{M}\sum_{\forall i'\neq i}g_{i',k,c,t}p_{m,i',c,t}+\sigma^2}\right),$$

which give the following set of equations:

$$w_{m,i,c,t}=\min_{k}\left(\frac{p_{m,i,c,t}\left[g_{i,k,c,t}+(1-\phi_{m,k})\hat{M}\right]}{\sum_{m=1}^{M}\sum_{\forall i'\neq i}g_{i',k,c,t}p_{m,i',c,t}+\sigma^2}\right)+1,\ \forall\,m,i,c,t.$$
$$(3.14)$$

and the following inequality set becomes valid:

$$\frac{p_{m,i,c,t}[g_{i,k,c,t} + (1 - \phi_{m,k})\hat{M}]}{\sum_{m=1}^{M} \sum_{\forall i' \neq i} g_{i',k,c,t} p_{m,i',c,t} + \sigma^2} \geq w_{m,i,c,t} - 1, \ \forall \, m, i, k, c, t.$$

(3.15)

Therefore, constraint set $C11$ can be replaced by:

$$\prod_{i=1}^{S} \prod_{c=1}^{C} \prod_{t=1}^{T} w_{m,i,c,t} \geq \hat{R}_m^{min} \theta_m + (1 - \theta_m), \ \forall \, m,$$

(3.16)

and

$$\frac{p_{m,i,c,t} \left[g_{i,k,c,t} + (1 - \phi_{m,k}) \hat{M} \right]}{\sum_{m=1}^{M} \sum_{\forall i' \neq i} g_{i',k,c,t} p_{m,i',c,t} + \sigma^2} \geq w_{m,i,c,t} - 1, \ \forall \, m, i, k, c, t.$$

(3.17)

where $w_{m,i,c,t} \geq 1$.

For $C12$, the constraint set can be rewritten in the form:

$$\log_2 \left[\prod_{i=1}^{S} \prod_{c=1}^{C} \prod_{t=1}^{T} \left(1 + \max_{k} \frac{g_{i,k,c,t} p_{m,i,c,t} \phi_{m,k}}{\sum_{m=1}^{M} \sum_{\forall i' \neq i} g_{i',k,c,t} p_{m,i',c,t} + \sigma^2} \right) \right] \leq \frac{R_m^{max} F}{\Delta B \Delta T}, \ \forall \, m,$$

(3.18)

taking the exponent of 2 for both sides we get:

$$\prod_{i=1}^{S} \prod_{c=1}^{C} \prod_{t=1}^{T} \left(1 + \max_{k} \frac{g_{i,k,c,t} p_{m,i,c,t} \phi_{m,k}}{\sum_{m=1}^{M} \sum_{\forall i' \neq i} g_{i',k,c,t} p_{m,i',c,t} + \sigma^2} \right) \leq \hat{R}_m^{max}, \ \forall \, m, \quad (3.19)$$

where $\hat{R}_m^{max} = 2^{\frac{R_m^{max} F}{\Delta B \Delta T}}$. Then we introduce the auxiliary variables $u_{m,i,c,t}$ for the terms

$$\left(1 + \max_{k} \frac{g_{i,k,c,t} p_{m,i,c,t} \phi_{m,k}}{\sum_{m=1}^{M} \sum_{\forall i' \neq i} g_{i',k,c,t} p_{m,i',c,t} + \sigma^2} \right),$$

which gives the following set of inequalities:

$$u_{m,i,c,t} = 1 + \max_{k} \left(\frac{g_{i,k,c,t} p_{m,i,c,t} \phi_{m,k}}{\sum_{m=1}^{M} \sum_{\forall i' \neq i} g_{i',k,c,t} p_{m,i',c,t} + \sigma^2} \right), \quad (3.20)$$
$$\forall \, m, i, c, t,$$

and the following inequality set becomes valid:

$$\frac{g_{i,k,c,t} p_{m,i,c,t} \phi_{m,k}}{\sum_{m=1}^{M} \sum_{\forall i' \neq i} g_{i',k,c,t} p_{m,i',c,t} + \sigma^2} \leq u_{m,i,c,t} - 1, \quad \forall \, m, i, k, c, t.$$
$$(3.21)$$

Therefore, the constraint $C12$ can be replaced by:

$$\prod_{i=1}^{S} \prod_{c=1}^{C} \prod_{t=1}^{T} u_{m,i,c,t} \leq \hat{R}_{m}^{max} \; \forall \, m, \qquad (3.22)$$

and

$$\frac{g_{i,k,c,t} p_{m,i,c,t} \phi_{m,k}}{\sum_{m=1}^{M} \sum_{\forall i' \neq i} g_{i',k,c,t} p_{m,i',c,t} + \sigma^2} \leq u_{m,i,c,t} - 1, \quad \forall \, m, i, k, c, t.$$
$$(3.23)$$

where $u_{m,i,c,t} \geq 1$. The new constraints given by (3.16), (3.17), (3.22) and (3.23) are all polynomials, where the ones given by (3.17) and (3.23) are second-degree polynomial (quadratic). Therefore, replacing the constraint sets C11 and C12 in $\mathcal{HAP}^{Eff}$ with (3.16), (3.17), (3.22) and (3.23) gives a mixed binary polynomial constraint program (MBPCP). Section 3.5 shows how this is further reduced to a mixed binary quadratically constrained program (MBQCP).

3.5 Reduction of the Formulation to a Mixed Binary Quadratic Constrained Program

Any MBPCP optimization problem can be reduced to a MBQCP by the introduction of auxiliary variables and constraints to reduce all polynomial degrees to 2. For example, a cubic polynomial term $x_1 x_2 x_3$ could be modeled as $x_1 X_{23}$ with $X_{23} = x_2 x_3$. Using this simple reformulation technique, the polynomial constraints obtained in the previous section can be converted to mixed binary quadratic constraints by replacing (3.16) by the following:

$$w_{m,(1)} W_{m,1} \geq \hat{R}_m^{min} \theta_m + (1 - \theta_m), \quad \forall\, m, \tag{3.24a}$$

$$W_{m,j} = w_{m,(j+1)} W_{m,j+1}, \quad \forall j = 1, 2, ..., n - 3, \ \forall\, m, \tag{3.24b}$$

$$W_{m,(n-2)} = w_{m,(n-1)} w_{m,(n)}, \quad \forall\, m, \tag{3.24c}$$

where $n = S.C.T$ and $j = (i - 1).C.T + (c - 1).T + t$ for the set of variables $W_{m,j}$ while for $w_{m,(j)}$, $j \equiv (i, c, t)$. Equality constraints can be replaced by inequality constraints to give:

$$w_{m,(1)} W_{m,1} \geq \hat{R}_m^{min} \theta_m + (1 - \theta_m), \quad \forall\, m, \tag{3.25a}$$

$$W_{m,j} \leq w_{m,(j+1)} W_{m,j+1}, \quad \forall j = 1, 2, ..., n - 3, \ \forall\, m, \tag{3.25b}$$

$$W_{m,j} \geq w_{m,(j+1)} W_{m,j+1}, \quad \forall j = 1, 2, ..., n - 3, \ \forall\, m, \tag{3.25c}$$

$$W_{m,n-2} \leq w_{m,(n-1)} w_{m,(n)}, \quad \forall\, m, \tag{3.25d}$$

$$W_{m,n-2} \geq w_{m,(n-1)} w_{m,(n)}, \quad \forall\, m, \tag{3.25e}$$

These sets replace the set of M constraints in (3.16) with $3M + 2M.(S.C.T - 3)$ quadratic constraints and adds $M \times (S.C.T - 2)$ new variables $W_{m,j}$. Similarly, the constraint set in (3.22) can be replaced by:

$$u_{m,(1)} U_{m,1} \leq \hat{R}_m^{max} \ \forall\, m, \tag{3.26a}$$

$$U_{m,j} \leq u_{m,(j+1)} U_{m,j+1} \ \forall j = 1, 2, ..., n - 3, \ \forall\, m \tag{3.26b}$$

$$U_{m,j} \geq u_{m,(j+1)} U_{m,j+1} \ \forall j = 1, 2, ..., n - 3, \ \forall\, m, \tag{3.26c}$$

$$U_{m,n-2} \leq u_{m,(n-1)} u_{m,(n)} \ \forall\, m, \tag{3.26d}$$

$$U_{m,n-2} \geq u_{m,(n-1)} u_{m,(n)} \ \forall\, m. \tag{3.26e}$$

Again, this replaces the M constraints in (3.22) with $3M + 2M.(S.C.T - 3)$ quadratic constraints and adds $M \times (S.C.T - 2)$ new variables $U_{m,j}$.

The optimization problem is now an MBQCP given by:

$$\max_{\substack{\phi_{m,k}, \theta_m, y_{m,i,c,t}, u_{m,i,c,t}, \\ U_{m,j}, w_{m,i,c,t}, W_{m,j}, p_{m,i,c,t}}} \sum_{m=1}^{M} \sum_{k=1}^{K} \rho_{m,k} \phi_{m,k} \qquad (\mathcal{HAP}_{MBQCP}^{Eff})$$

s.t.

$\overline{C1} : \phi_{m,k} \leq \lambda_{m,k}, \ \forall \, m, k$

$\overline{C2} : \displaystyle\sum_{m=1}^{M} y_{m,i,c,t} \leq 1, \ \forall i, c, t$

$\overline{C3} : \displaystyle\sum_{i=1}^{S} \sum_{c=1}^{C} \sum_{t=1}^{T} y_{m,i,c,t} \geq \phi_{m,k}, \ \forall m, k$

$\overline{C4} : y_{m,i,c,t} \leq \displaystyle\sum_{k=1}^{K} \phi_{m,k}, \ \forall \, m, i, c, t$

$\overline{C5} : P_{PF}^{Total} y_{m,i,c,t} \geq p_{m,i,c,t}, \ \forall m, i, c, t$

$\overline{C6} : \displaystyle\sum_{m=1}^{M} \sum_{i=1}^{S} \sum_{c=1}^{C} p_{m,i,c,t} \leq P_{PF}^{Total}, \ \forall t$

$\overline{C7} : p_{m,i,c,t} \geq 0, \ \forall \, m, i, c, t$

$\overline{C8} : \dfrac{g_{i,k,c,t} p_{m,i,c,t} + (1 - \phi_{m,k}) \hat{M}}{\sum_{m=1}^{M} \sum_{i' \neq i} g_{i',k,c,t} p_{m,i',c,t} + \sigma^2} \geq y_{m,i,c,t} \gamma_{m,i}^{th}, \ \forall m, i, k, c, t$

$\overline{C9} : \displaystyle\sum_{i=1}^{S} \sum_{c=1}^{C} \sum_{t=1}^{T} y_{m,i,c,t} \leq SCT\theta_m, \ \forall \, m$

$\overline{C10} : \displaystyle\sum_{i=1}^{S} \sum_{c=1}^{C} \sum_{t=1}^{T} y_{m,i,c,t} \geq \theta_m, \ \forall \, m$

$\overline{Q1a} : w_{m,(1)} W_{m,1} \geq \hat{R}_m^{min} \theta_m + (1 - \theta_m), \ \forall \, m$

$$\overline{Q1b} : W_{m,j} \leq w_{m,(j+1)} W_{m,j+1} \; \forall j = 1, 2, ..., n - 3, \; \forall m$$

$$\overline{Q1c} : W_{m,j} \geq w_{m,(j+1)} W_{m,j+1} \; \forall j = 1, 2, ..., n - 3, \; \forall m$$

$$\overline{Q1d} : W_{m,n-2} \leq w_{m,(n-1)} w_{m,(n)}, \; \forall m$$

$$\overline{Q1e} : W_{m,n-2} \geq w_{m,(n-1)} w_{m,(n)}, \; \forall m$$

$$\overline{Q2} : \frac{p_{m,i,c,t} \left[g_{i,k,c,t} + (1 - \phi_{m,k}) \hat{M} \right]}{\sum_{m=1}^{M} \sum_{\forall i' \neq i} g_{i',k,c,t} p_{m,i',c,t} + \sigma^2} \geq w_{m,i,c,t} - 1, \; \forall m, i, k, c, t$$

$$\overline{Q3a} : u_{m,(1)} U_{m,1} \leq \hat{R}_m^{max}, \; \forall m$$

$$\overline{Q3b} : U_{m,j} \leq u_{m,(j+1)} U_{m,j+1} \; \forall j = 1, 2, ..., n - 3, \; \forall m$$

$$\overline{Q3c} : U_{m,j} \geq u_{m,(j+1)} U_{m,j+1} \; \forall j = 1, 2, ..., n - 3, \; \forall m$$

$$\overline{Q3d} : U_{m,n-2} \leq u_{m,(n-1)} u_{m,(n)} \; \forall m$$

$$\overline{Q3e} : U_{m,n-2} \geq u_{m,(n-1)} u_{m,(n)} \; \forall m$$

$$\overline{Q4} : \frac{g_{i,k,c,t} p_{m,i,c,t} \phi_{m,k}}{\sum_{m=1}^{M} \sum_{\forall i' \neq i} g_{i',k,c,t} p_{m,i',c,t} + \sigma^2} \leq u_{m,i,c,t} - 1 \; \forall m, i, k, c, t$$

$$\phi_{m,k}, \theta_m, y_{m,i,c,t} \in \{0, 1\} \; \forall m, i, k, c, t$$

$$0 \leq p_{m,i,c,t} \leq P_{PF}^{Tot}, \; 1 \leq w_{m,i,c,t} \leq \hat{R}_m^{max},$$

$$1 \leq W_{m,j} \leq \hat{R}_m^{max} \; 1 \leq u_{m,i,c,t} \leq \hat{R}_m^{max},$$

$$0 \leq U_{m,j} \leq \hat{R}_m^{max}, \; \forall m, i, c, t.$$

3.6 Comparison of the Formulation Sizes with the Aid of a Numerical Example

In this section, we illustrate the differences in the sizes of the formulations (OP1) in [4] and $\mathcal{HAP}_{MBQCP}^{Eff}$. We provide P-Prob's formulation (OP1) here for reference and comparison:

$$\max_{z_{m,i,k,c,t}, p_{m,i,c,t}} \sum_{m=1}^{M} \sum_{i=1}^{S} \sum_{k=1}^{K} \sum_{c=1}^{C} \sum_{t=1}^{T} z_{m,i,k,c,t} \qquad \text{(OP1)}$$

s.t.

$$D1 : z_{m,i,k,c,t} \leq \lambda_{m,i,k}, \quad \forall m, i, k, c, t$$

$$D2 : z_{m,i,k,c,t} + z_{m,i,k',c',t'} \leq 1 + z_{m,i,k,c',t'}, \forall m, i;$$
$$\forall k, k' : k \neq k'; \forall c, c' : c \neq c'; \forall t, t' : t \neq t'$$

$$D3 : z_{m,i,k,c,t} \geq \frac{A_m - \Omega}{\frac{g_{i,k,c,t}P_{m,i,c,t}}{\sum_{m=1}^{M} \sum_{\substack{\forall i' \in S \\ i' \neq i}} g_{i',k,c,t}P_{m,i',c,t} + \sigma^2} - \Omega}, \quad \forall m, i, k, c, t$$

$$D4 : z_{m,i,k,c,t} \leq \frac{B_m \left(\sum_{m=1}^{M} \sum_{\substack{\forall i' \in S \\ i' \neq i}} g_{i',k,c,t}P_{m,i',c,t} + \sigma^2 \right)}{g_{i,k,c,t}P_{m,i,c,t}}, \forall m, i, k, c, t$$

$$D6 : \sum_{c=1}^{C} \sum_{t=1}^{T} z_{m,i,k,c,t} \geq z_{m,i,k,c,t}y_m^{min}, \quad \forall m, i, k, c, t$$

$$D7 : \sum_{c=1}^{C} \sum_{t=1}^{T} z_{m,i,k,c,t} \leq y_m^{max}, \quad \forall m, i, k$$

$$D8 : z_{m,i,k,c,t} \in \{0, 1\}, \quad \forall m, i, k, c, t$$

$$D9 : \sum_{m=1}^{M} \sum_{i=1}^{S} \sum_{c=1}^{C} p_{m,i,c,t} \leq P_{PF}^{total}, \quad \forall t$$

$$D10 : p_{m,i,c,t} \leq P_{PF}^{total} \sum_{k=1}^{k=K} z_{m,i,k,c,t}, \quad \forall m, i, c, t$$

$$D11 : p_{m,i,c,t} \geq 0, \quad \forall m, i, c, t.$$

For the interpretation of the constraints in (OP1), we refer the reader to [2–4]. Considering (OP1) in [4] first, we see that the number of variables are as follows:

- The number of binary variables, $z_{m,i,k,c,t}$, is the product $MSKCT$ and
- the number of continuous variables, $p_{m,i,c,t}$, is $MSCT$,

- hence giving a total number of variables

$$VN_{OP1} = MSKCT + MSCT. \qquad (3.27)$$

The number of constraints (excluding bounds and integrality constraints) in each constraint set for (OP1) in [4] are as follows:

- constraint set $D1$ comprises $MSKCT$ constraints,
- constraint set $D2$ comprises $MSKCT[CT-1][K-1]$ constraints,
- constraint set $D3$ comprises $MSKCT$ constraints,
- constraint set $D4$ comprises $MSKCT$ constraints,
- constraint set $D5$ comprises $MSKCT[M-1][K-1]$ constraints,
- constraint set $D6$ comprises $MSKCT$ constraints,
- constraint set $D7$ comprises MSK constraints,
- constraint set $D9$ comprises T constraints,
- constraint set $D10$ comprises $MSCT$ constraints,

which all add up to

$$CN_{OP1} = MSKCT[CT-1][K-1] +$$
$$MSKCT[M-1][K-1] + 4MSKCT + MSK + MSCT + T.$$
$$(3.28)$$

For the formulation $\mathcal{HAP}^{Eff}_{MBQCP}$, we have the following numbers of variables:

- The numbers of binary variables $\phi_{m,k}, \theta_m, y_{m,i,c,t}$ are the MK, M and $MSCT$, respectively, giving a total number of binary variables $MK + M + MSCT$.
- The number of continuous variables:
  - $p_{m,i,c,t}$ are $MSCT$,
  - $u_{m,i,c,t}$ are $MSCT$,
  - $w_{m,i,c,t}$ are $MSCT$
  - $U_{m,j}$ are $M[SCT-2]$, and
  - $W_{m,j}$ are $M[SCT-2]$.

  all adding up to $3MSCT + 2M[SCT-2]$ continuous variables.

The number of binary and continuous variables add up to:

$$VN_{\mathcal{HAP}^{Eff}_{MBQCP}} = 4MSCT + 2M\left[SCT - 2\right] + MK + M. \quad (3.29)$$

The number of constraints (excluding variable bounds and binary constraints) in each constraint set for $\mathcal{HAP}^{Eff}_{MBQCP}$ are as follows:

- Constraint set $\overline{C1}$ consists of MK constraints,
- Constraint set $\overline{C2}$ consists of SCT constraints,
- Constraint set $\overline{C3}$ consists of MK constraints,
- Constraint set $\overline{C4}$ consists of $MSCT$ constraints,
- Constraint set $\overline{C5}$ consists of $MSCT$ constraints,
- Constraint set $\overline{C6}$ consists of T constraints,
- Constraint set $\overline{C8}$ consists of $MSKCT$ constraints,
- Constraint set $\overline{C9}$ consists of M constraints,
- Constraint set $\overline{C10}$ consists of M constraints,
- Constraint set $\overline{Q1a}$ consists of M constraints,
- Constraint set $\overline{Q1b}$ consists of $M\left[SCT - 3\right]$ constraints,
- Constraint set $\overline{Q1c}$ consists of $M\left[SCT - 3\right]$ constraints,
- Constraint set $\overline{Q1d}$ consists of M constraints,
- Constraint set $\overline{Q1e}$ consists of M constraints,
- Constraint set $\overline{Q2}$ consists of $MSKCT$ constraints,
- Constraint set $\overline{Q3a}$ consists of M constraints,
- Constraint set $\overline{Q3b}$ consists of $M\left[SCT - 3\right]$ constraints,
- Constraint set $\overline{Q3c}$ consists of $M\left[SCT - 3\right]$ constraints,
- Constraint set $\overline{Q3d}$ consists of M constraints,
- Constraint set $\overline{Q3e}$ consists of M constraints,
- Constraint set $\overline{Q4}$ consists of $MSKCT$ constraints,

which all add up to

$$CN_{\mathcal{HAP}^{Eff}_{MBQCP}} = 2MK + SCT + 2MSCT + T + 8M$$
$$+ 4M\left[SCT - 3\right] + 3MSKCT. \quad (3.30)$$

Finally, both formulations (OP1) in [4] and $\mathcal{HAP}^{Eff}_{MBQCP}$ consist of bilinear terms. By counting the bilinear terms in (OP1) obtained from constraints sets $D3$ and $D4$, we obtain:

$$N^{BiL}_{\mathcal{HAP}^{Lagrange}_2} = M^2 S^2 KCT + MSKCT. \tag{3.31}$$

Also, by counting the bilinear terms in constraint sets $\overline{C8}, \overline{Q1a}, \overline{Q2}, \overline{Q1b}$, $\overline{Q1c}$, $\overline{Q1d}, \overline{Q1e}, \overline{Q3a}$, $\overline{Q3b}, \overline{Q3c}, \overline{Q3d}, \overline{Q3e}$ and $\overline{Q4}$ of $\mathcal{HAP}^{Eff}_{MBQCP}$, we obtain:

$$N^{BiL}_{\mathcal{HAP}^{Eff}_{MBQCP}} = M^2 S (S - 1) KCT +$$
$$2MSKCT [1 + M (S - 1)] + 4M [SCT - 3] + 6M. \tag{3.32}$$

We graphically illustrate a comparison of formulation efficiency for the two formulations (OP1) and $\mathcal{HAP}^{Eff}_{MBQCP}$ in Figures 3.3–3.7. In these figures, we compare the number of binary variables, continuous variables, total number of variables, number of constraints and number of bilinear terms for both formulations. We refer to the indices m, i, k, c and t as the problem "dimensions". Therefore, there are five dimensions for the problem in both formulations, which are the number of multicast sessions, the number of HAP antennas on-board, the number of users in the service area, the number of sub-channels and the number of time slots respectively. We vary the dimensions of the problem as follows:

- The number of multicast sessions, M, is varied in the range 1–250,
- the number of antennas on-board, S, is varied in the range 1–20,
- the number of users, K, in the service area is varied in the range 1–500,
- the number of available sub-channels, C, is varied in the range 1–32 and
- the number of available sub-channels, T, is varied in the range 1–24.

Figures 3.3–3.7 comprise five plots in each of which one dimension is varied within its ranges mentioned above and the others are kept

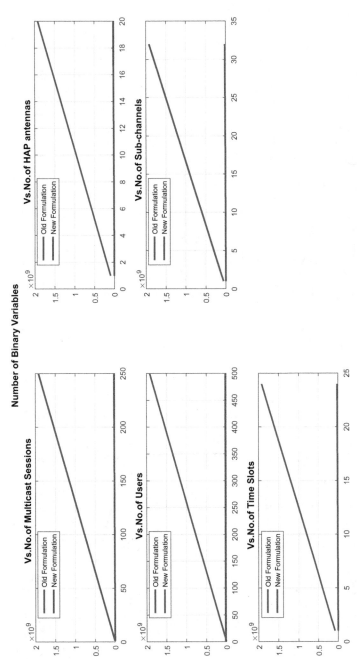

Figure 3.3   Illustration of the number of binary variables versus the different problem dimensions for (OP1) in [4] (old formulation) and $\mathcal{HAP}_{MBQCP}^{Eff}$ (new formulation).

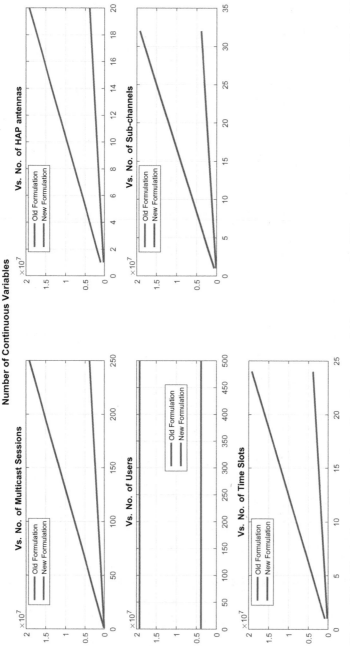

Figure 3.4 Illustration of the number of continuous variables versus the different problem dimensions for (OP1) in [4] (old formulation) and $\mathcal{H}AP_{MBQCP}^{Eff}$ (new formulation).

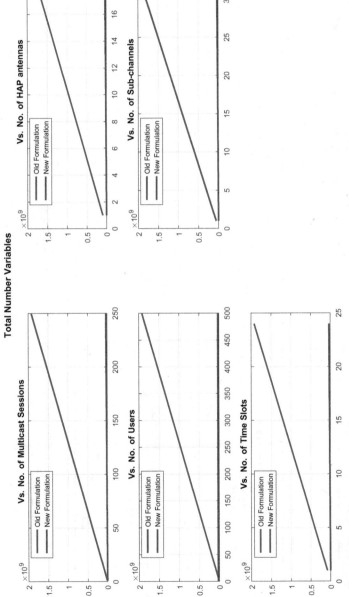

Figure 3.5 Illustration of the total number of variables versus the different problem dimensions for (OP1) in [4] (old formulation) and $\mathcal{HAP}^{Eff}_{MBQCP}$ (new formulation).

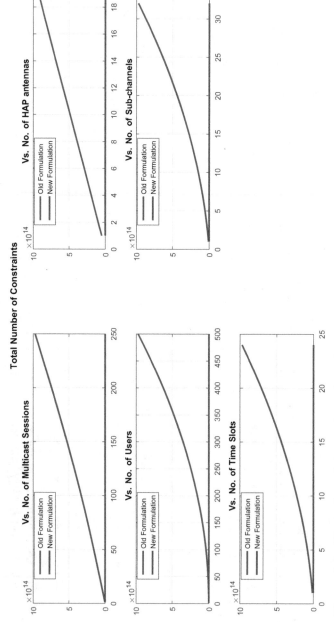

Figure 3.6 Illustration of the total number of constraints versus the different problem dimensions for (OP1) in [4] (old formulation) and $\mathcal{HAP}^{Eff}_{MBQCP}$ (new formulation).

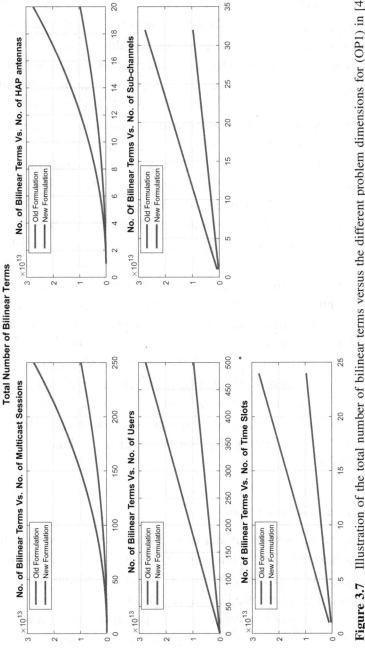

Figure 3.7 Illustration of the total number of bilinear terms versus the different problem dimensions for (OP1) in [4] (old formulation) and $\mathcal{HAP}^{Eff}_{MBQCP}$ (new formulation).

fixed at values equal to their maximums in their respective ranges. The results in Figure 3.3 show that the number of binary variables for $\mathcal{HAP}^{Eff}_{MBQCP}$ is way lower than those in (OP1). On the other hand, in Figure 3.4, the number of continuous variables $\mathcal{HAP}^{Eff}_{MBQCP}$ are almost 4 times those of (OP1) for the worst case. However, by looking at both Figures 3.3 and 3.4, we can see that the number of continuous variables in both formulations are much lower than the binary variables which makes the total number of variables in Figure 3.5 almost equivalent to the total number of binary variables. Moreover, it is well known that when there are both binary variables and continuous variables in a problem, the binary variables are the main cause of algorithmic complexity involved in solving the problem. Therefore, comparing the numbers of continuous and binary variables in both formulations, we see that $\mathcal{HAP}^{Eff}_{MBQCP}$ has a much lower complexity compared to (OP1).

Taking a look at the number of total constraints in Figure 3.6, we see that the number of constraints in formulation $\mathcal{HAP}^{Eff}_{MBQCP}$ is far lower than (OP1). This comes at the cost of up to three times larger number of bilinear terms, in the worst case, for $\mathcal{HAP}^{Eff}_{MBQCP}$ in all dimensions as shown in Figure 3.7. Notice the similar behaviors for both $\mathcal{HAP}^{Eff}_{MBQCP}$ and (OP1) in Figure 3.7 for each dimension. For the dimensions of the number of multicast sessions, m, and the number of HAP antenna onboard, i, the number of bilinear terms for both formulations grows quadratically. For the other three dimensions, the growth is linear.

3.7 Chapter Conclusion

In this chapter, the system model of E-Prob was presented, in which a user was allowed to join any session being transmitted in all the neighboring cells, whereas in P-Prob a user can only join a subset of the sessions being transmitted in the cell it resides. Moreover, E-Prob allows a multicast group to receive transmission of a session on more than one antenna simultaneously, which is a more flexible option that is not available in the primary model. Also, the extended system model takes into account heterogeneous user and session priorities which are considered to be homogeneous in P-Prob. The chapter focused on a derivation of a formulation whose new variable definitions enabled us to obtain a much smaller formulation than that obtained earlier for P-Prob. As explained in Section 2.2, the computational effort and memory requirements for solving an optimization problem is directly dependent on the size of the formulated problem, besides other aspects. Succeeding in reducing the problem size significantly, as we have done and showed in this chapter, is an important achievement for multicasting AC-RRA over HAP, especially when the radio channel gains are changing quickly. In the next two chapters, we will present the solution method, based on a branch and bound framework using McCormick underestimators, for the derived formulation $\mathcal{HAP}^{Eff}_{MBQCP}$. Each chapter will focus on some particular components of the solution framework and present their numerical experiments' results.

4

Proposed Solution Method: Branching Schemes and a Presolving Linearization-Based Reformulation

This chapter and Chapter 5 explain how formulation $\mathcal{HAP}^{Eff}_{MBQCP}$ is solved. An approach similar to [57] and [56] is used in which an outer approximation is generated by linear underestimation of the non-convex quadratic constraints to relax the problem's feasible region. The problem becomes an MBLP and hence an LP solver can be used in a branch-and-cut algorithm to solve $\mathcal{HAP}^{Eff}_{MBQCP}$. The branch-and-bound (BnB) algorithm recursively splits the problem into smaller subproblems, thereby creating a branching tree and implicitly enumerating all potential solutions. At each subproblem, domain propagation is performed to exclude further values from the variables' domains, and a relaxation is solved to achieve an upper (dual) bound. The relaxation is then strengthened by adding further valid constraints, which cut off the optimum of the relaxation. Primal heuristics are integrated in the BnB procedure to improve the lower (primal) bound. The solver used for the experiments is *Solving Constraint Integer Programs* (SCIP) which is capable of solving a non-convex *mixed integer quadratically constraint program* (MIQCP) to optimality in finite time [59]. The interdependencies between the algorithmic components of SCIP solver are shown in Figure 4.1. An explanation for the solution algorithm's components used to solve the $\mathcal{HAP}^{Eff}_{MBQCP}$, and the corresponding experiments conducted, are presented in this chapter and Chapter 5. The components are:

- Presolving
- Branching

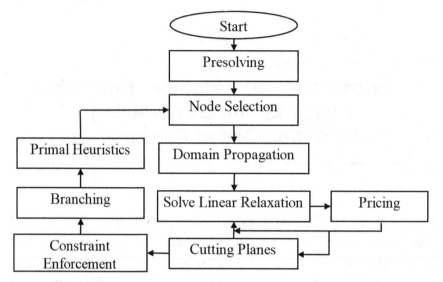

Figure 4.1   Flowchart illustrating the interrelation between the SCIP solver components used for solving the HAP multicasting AC-RRA problem.

- Separating Cuts
- Domain Propagation
- Primal Heuristics

The two components considered in this chapter are presolving and branching.

4.1 A Presolving Linearization for a Particular Quadratic Constraint Set of the Formulation

Presolving is a set of operations invoked before the branch-and-bound algorithm to transform the problem instance to an easier instance to solve. In this section, for the presolving phase of the solution approach of $\mathcal{HAP}^{Eff}_{MBQCP}$, we consider one of the reformulations in [57] which is a linear representation for bilinear terms that are a product of a binary variable $\ddot{x}$ with a continuous linear term i.e. $\ddot{x} \sum_{\forall i} a_i \ddot{y}_i$. This type of reformulation is applicable to the constraint set $\overline{C8}$ in $\mathcal{HAP}^{Eff}_{MBQCP}$ where the terms that consist of the product of binary variables and linear terms are $y_{m,i,c,t} \sum_{m=1}^{M} \sum_{i=1}^{S} g_{i',k,c,t} p_{m,i',c,t}$. The product is replaced by the auxiliary variable z and the linear constraints:

$$\tilde{p}^L y_{m,i,c,t} \leq z \leq \tilde{p}^U y_{m,i,c,t}, \tag{4.1a}$$

$$\sum_{m=1}^{M} \sum_{\forall i' \neq i} g_{i',k,c,t} p_{m,i',c,t} - \tilde{p}^L \left(1 - y_{m,i,c,t}\right) \leq z$$

$$\leq \sum_{m=1}^{M} \sum_{\forall i' \neq i} g_{i',k,c,t} p_{m,i',c,t} - \tilde{p}^U \left(1 - y_{m,i,c,t}\right), \tag{4.1b}$$

where

$$\tilde{p}^L = \sum_{m=1}^{M} \sum_{\forall i' \neq i} g_{i',k,c,t} \tilde{p}^L_{m,i',c,t}, \tag{4.2a}$$

$$\tilde{p}^U = \sum_{m=1}^{M} \sum_{\forall i' \neq i} g_{i',k,c,t} \tilde{p}^U_{m,i',c,t}, \tag{4.2b}$$

given the local lower and upper bounds $\tilde{p}^L_{m,i',c,t}$ and $\tilde{p}^U_{m,i',c,t}$, respectively, of the decision variable $p_{m,i',c,t}$. This reformulation linearizes the constraint set $\overline{C8}$ at the expense of introducing one continuous variable for each constraint in the set, and four linear constraints for each quadratic constraint in $\overline{C8}$. In Section 4.4, the algorithmic performance criteria, with and without, the presolving linearization reformulation explained in this section, for different number of presolving rounds, are presented.

After the presolving phase, the BnB algorithm is invoked. Any reference for $\mathcal{HAP}^{Eff}_{MBQCP}$ in the rest of this section refers to the instance *after* going through the presolving phase.

4.2 Branch and Bound-Based Solution Framework

The branch and bound scheme is a general framework used in solving non-convex problems, which include MBLPs and MBQLPs, to divide it into smaller problems that can be solved (conquered) and hence is a divide and conquer algorithm [60]. The best local solution across all the subproblems, which are referred to as nodes, is the global solution of the entire problem. *Branching* is basically the splitting of a subproblem into two or more nodes. Since the discrete variables that we have in $\mathcal{HAP}_{MBQCP}^{Eff}$ are binary by nature, binary branching is the only choice, i.e. no more than two children nodes for any node in the tree. The *root* node is the whole problem $\mathcal{HAP}_{MBQCP}^{Eff}$ before division, while the rest of the nodes are smaller subproblems.

The *bounding* step avoids complete enumeration of potential solutions of the problem. The better the dual $\ddot{c}_{dual}$ and primal $\ddot{c}_{primal}$ bounds are, the more effective the bounding process in excluding subproblems from solving. The dual bound is found by solving the relaxation $\mathcal{Q}_{relax}$ of a node subproblem $\mathcal{Q}$. The relaxation $\mathcal{Q}_{relax}$ for $\mathcal{HAP}_{MBQCP}^{Eff}$ is obtained by replacing all the bilinear terms individually by McCormick linear under-estimators [57], and by relaxing all the binary variables into the continuous domain [0,1]. Algorithm 1 illustrates the main procedures of a BnB framework. To simplify the notation in the rest of our discussion, and wherever specific reference to certain variables in our formulation $\mathcal{HAP}_{MBQCP}^{Eff}$ is not required, all the decision variables in $\mathcal{HAP}_{MBQCP}^{Eff}$ are represented by the decision vector $\ddot{x}$. Furthermore, an arbitrary decision variable is referred to as $\ddot{x}_j$, where $j \in \tilde{N}$ for any variable and if the variable is binary then additionally $j \in \mathcal{B}$, where $\tilde{N}$ is the set of all decision variables and $\mathcal{B}$ is the set of all binary decision variables in $\mathcal{HAP}_{MBQCP}^{Eff}$.

The input to the algorithm is a presolved instance of $\mathcal{HAP}_{MBQCP}^{Eff}$ which resembles the root node. If the instance is feasible, then the output of the algorithm is the global optimal solution $\ddot{x}_{opt}^{\mathcal{HAP}_{MBQCP}^{Eff}}$ and the corresponding objective function value $\ddot{c}_{opt}^{\mathcal{HAP}_{MBQCP}^{Eff}}$, otherwise the

1: **Input**: Maximization of an instance of $\mathcal{HAP}^{Eff}_{MBQCP}$

2: **Output**: Optimal solution $\ddot{\mathbf{x}}^{\mathcal{HAP}^{Eff}_{MBQCP}}_{opt}$ with objective function value $\ddot{c}^{\mathcal{HAP}^{Eff}_{MBQCP}}_{opt}$

    or conclusion that $\mathcal{HAP}^{Eff}_{MBQCP}$ has no solution by $\ddot{c}^{\mathcal{HAP}^{Eff}_{MBQCP}}_{opt} = -\infty$

    <u>**Initialize**:</u>

3: $\mathcal{Q} \leftarrow \mathcal{HAP}^{Eff}_{MBQCP}$

4: $\ddot{\mathcal{L}} = \{\mathcal{Q}\}$

5: $\ddot{c}_{primal} = -\infty$

    <u>**Abort**:</u>

6: **if** $\ddot{\mathcal{L}} = \emptyset$ **then**

7:      $\ddot{\mathbf{x}}^{\mathcal{HAP}^{Eff}_{MBQCP}}_{opt} \leftarrow \ddot{\mathbf{x}}_{BFS}$

8:      $\ddot{c}^{\mathcal{HAP}^{Eff}_{MBQCP}}_{opt} \leftarrow \ddot{c}_{primal}$

9:      **STOP**

10: **end if**

    <u>**Select**:</u>

11: Choose $\mathcal{Q} \in \ddot{\mathcal{L}}$ and

12: $\ddot{\mathcal{L}} \leftarrow \ddot{\mathcal{L}} \setminus \{\mathcal{Q}\}$

    <u>**Solve**:</u>

13: Solve the linear relaxation $\mathcal{Q}_{relax}$ after applying McCormick under-estimators to all bilinear terms of $\mathcal{HAP}^{Eff}_{MBQCP}$.

14: **if** $\mathcal{Q}_{relax} = \emptyset$ **then**

15:     $\ddot{c}_{dual} \leftarrow -\infty$

16: **else**

17:     let $\ddot{\mathbf{x}}_{relax}$ be the optimal solution of $\mathcal{Q}_{relax}$ and $\ddot{c}_{dual}$ its objective function value.

18: **end if**

    <u>**Bound**:</u>

19: **if** $\ddot{c}_{dual} \leq \ddot{c}_{primal}$ **then**

20:     Prune node $\mathcal{Q}$

21:     **goto** step (6)

22: **end if**

    <u>**Feasibility Check**:</u>

23: **if** $\ddot{\mathbf{x}}_{relax}$ is feasible for $\mathcal{HAP}^{Eff}_{MBQCP}$ **then**

24:     $\ddot{\mathbf{x}}_{BFS} \leftarrow \ddot{\mathbf{x}}_{relax}$

25:     $\ddot{c}_{primal} \leftarrow \ddot{c}_{dual}$

26:     **goto** step (6)

27: **end if**

    <u>**Branch**:</u>

28: Divide $\mathcal{Q}$ into two subproblems $\mathcal{Q} = \mathcal{Q}_0 \cup \mathcal{Q}_1$

29: $\ddot{\mathcal{L}} \leftarrow \{\mathcal{Q}_0 \cup \mathcal{Q}_1\}$

30: **goto** step (6).

Algorithm 1: Branch-and-Bound Solution Framework for solving $\mathcal{HAP}^{Eff}_{MBQCP}$

algorithm concludes that the instance is infeasible. The algorithm is initialized by assigning the root node $\mathcal{HAP}^{Eff}_{MBQCP}$ into the empty node queue $\tilde{\mathcal{L}}$. The **Abort** procedure is invoked when the node queue is empty to return the best feasible solution found so far $\ddot{x}_{BFS}$ and its corresponding objective function value $\ddot{c}_{primal}$. If the node queue still has further unprocessed nodes, the **Select** procedure is invoked to choose a node $\mathcal{Q}$ depending on a *node selection criterion* before it gets removed from the queue. The relaxation of the selected node $\mathcal{Q}_{relax}$ is solved using the simplex algorithm [61] after applying McCormick underestimators to outer-approximate the non-convex quadratic constraints of $\mathcal{HAP}^{Eff}_{MBQCP}$. If $\mathcal{Q}_{relax}$ is found infeasible then $\ddot{c}_{dual}$ is assigned the smallest possible value (theoretically $-\infty$) to insure that the node gets pruned in the **Bound**. Otherwise $\ddot{x}_{relax}$ becomes the solution of $\mathcal{Q}_{relax}$ and $\ddot{c}_{dual}$ is its corresponding objective function value.

The **Bound** procedure is responsible for pruning branches from the search tree whose descendant nodes are guaranteed not to include any solutions better than the currently available best feasible solution (incumbent) $\ddot{x}_{BFS}$. This is known by doing a simple comparison between the obtained $\ddot{c}_{dual}$ from the **Solve** procedure and the objective function value $\ddot{c}_{primal}$ for the incumbent. In a maximization problem, like $\mathcal{HAP}^{Eff}_{MBQCP}$, if the dual (upper) bound is lower than the primal (lower) bound value, this is an indication that any of the descendants of the node can never have any better feasible solutions. If the node gets pruned, the algorithm goes back to the **Abort** procedure to check if there are any nodes left in the queue $\tilde{\mathcal{L}}$. If no pruning occurs, the **Feasibility Check** procedure is invoked and sets the solution $\ddot{x}_{relax}$ of the relaxed subproblem $\mathcal{Q}_{relax}$ as the solution of the $\mathcal{Q}$ itself only if the solution $\ddot{x}_{relax}$ is feasible to $\mathcal{Q}$. If $\ddot{x}_{relax}$ is not feasible to $\mathcal{Q}$, the **Branch** procedure then gets invoked to divide node $\mathcal{Q}$ into further nodes. This happens by selecting an appropriate variable to branch on. Since all the discrete variables in $\mathcal{HAP}^{Eff}_{MBQCP}$ are binary, then the branching is also binary. After branching takes place, the **Abort** procedure gets invoked to check whether there are any unprocessed nodes left in $\ddot{\mathcal{L}}$.

The node selection indicated by the **Select** procedure, the branching rules indicated by **Branch** and the relaxation whose solution is used in the **Bound** procedure all have a major impact on how early good feasible solutions can be found and how fast the dual bounds decreases. They all influence the **Bound** procedure which is expected to prune large parts of the BnB tree. The different branching rules used in the experiments conducted on $\mathcal{HAP}^{Eff}_{MBQCP}$ are discussed in Section 4.3.

4.3 Branching Techniques

Branching is the splitting of a node into two or more nodes by adding new upper and lower bounds on one of the variables which is called the *branching variable*. By reducing a variable's domain, the created children nodes have smaller feasible regions each, which helps reduce the work required to find feasible solutions better than the currently available best feasible solution $\ddot{x}_{BFS}$.

One advantage of using LP relaxation within BnB is in the branching process. Branching changes the created children subproblems from a parent node Q by introducing new upper or lower bound to one variable which preserves the dual feasibility of the solution obtained for Q_{relax}. This enables the use of dual simplex using the parent node solution as a warm-up start and hence the work done in solving Q_{relax} counts towards solving the relaxation of its children, which saves a lot of further computational work.

For $\mathcal{HAP}^{Eff}_{MBQCP}$, the only discrete variables are binary, and hence two nodes only are created by branching on binary variables. A score s_j^{branch} is calculated for each variable using equation [62]:

$$s_j^{branch} = max\left\{\ddot{q}_j^0, \bar{\epsilon}\right\} max\left\{\ddot{q}_j^1, \bar{\epsilon}\right\}, \qquad (4.3)$$

which measures the improvement in the dual bound by branching on the variable $\ddot{x}_j$ for $j \in \mathcal{B}$ where:

- $\mathcal{B}$ is the set of binary variables in $\mathcal{HAP}_{MBQCP}^{Eff}$,
- $\ddot{q}_j^0$ is a function that is directly dependent on and proportional to the dual-bound improvement $\ddot{\Delta}_j^0$ over the parent subproblem's relaxation $\mathcal{Q}_{relax}$ by setting $\ddot{x}_j = 0$,
- $\ddot{q}_j^1$ is a function that is directly dependent on and proportional to the dual-bound improvement $\ddot{\Delta}_j^1$ over the parent subproblem's relaxation $\mathcal{Q}_{relax}$ by setting $\ddot{x}_j = 1$,
- $\bar{\epsilon}$ which is a very small positive constant which is necessary to compare $\left(\ddot{\Delta}_j^0, \ddot{\Delta}_j^1\right)$ and $\left(\ddot{\Delta}_k^0, \ddot{\Delta}_k^1\right)$ and set by default to $\bar{\epsilon} = 10^{-6}$ in SCIP.

There are many different ways by which a branching variable can be selected, and of course, they can have different performances in bound improvement which are illustrated in the results provided in Section 4.4. The following branching schemes are considered for solving $\mathcal{HAP}_{MBQCP}^{Eff}$.

4.3.1 Random Branching

As the name indicates, there is nothing done in this technique except arbitrarily selecting any unfixed binary variable that violates the binary condition.

4.3.2 Most Infeasible Branching

This rule chooses the variable with the smallest tendency to be rounded either downwards or upwards. Hence, for binary variables with fractional values in the solution of $\mathcal{Q}_{relax}$, the one that is closest to 0.5 receives the highest score. The score function for a fractional binary variable is given as:

$$s_j^{branch} = min\left\{\ddot{x}_j^{relax}, 1 - \ddot{x}_j^{relax}\right\}, \quad j \in \mathcal{B}. \qquad (4.4)$$

4.3.3 Pseudocost Branching

This type of branching keeps a history for the average performance of each variable that has been branched on so far. This is measured as the average improvement in the bound for all the times the variable has been branched on. To obtain the variable scores, first the unit bound change for $\ddot{x}_j$ is found using

$$\varsigma_j^0 = \frac{\ddot{\Delta}^0}{\ddot{x}_j^{relax}} \quad and \quad \varsigma_j^1 = \frac{\ddot{\Delta}^1}{\left(1 - \ddot{x}_j^{relax}\right)}. \tag{4.5}$$

Let the aggregate unit bound changes be $\ddot{\sigma}_j^0$ and $\ddot{\sigma}_j^1$ over all nodes for which $\ddot{x}_j$ was selected for branching and the numbers of these nodes be η_j^0 and η_j^1 then the pseudocosts of $\ddot{x}^j$ are the averages:

$$\Psi_j^0 = \frac{\ddot{\sigma}_j^0}{\eta_j^0} \quad and \quad \Psi_j^1 = \frac{\ddot{\sigma}_j^1}{\eta_j^1}. \tag{4.6}$$

The score is then given as:

$$s_j^{branch} = max\left\{\ddot{x}_{relax}^j \Psi_j^0, \bar{\epsilon}\right\}.max\left\{\left(1 - \ddot{x}_{relax}^j\right)\Psi_j^1, \bar{\epsilon}\right\}. \tag{4.7}$$

During the course of Algorithm 1, a variable that has not yet been selected for branching is termed to be *uninitialized*, a term that will be used in subsequent subsections.

4.3.4 Strong Branching

This can be summarized as solving the linear relaxations that result from branching on each binary branching candidate and choosing the variable that gives the best bound improvement to branch on. It is hence expected that to obtain the optimal solution for a problem instance, the number of nodes required to be explored is going to be low but the number of simplex LP iterations is going to be too high which consumes a lot of processing time. The full set of branching candidates F_{cand} are all binary variables with fractional values. The entire set could be used in strong branching or only a subset $\overline{F} \subset F_{cand}$.

4.3.5 Hybrid Strong/Pseudocost Branching

Strong branching and pseudocost branching both have their advantages and draw-backs. For strong branching, as mentioned in Section 4.3.4, the number of nodes to be explored before the optimal solution is reached is expected to be low. However, the time and LP iterations expended could be too high. On the other hand, pseudocost branching is not expected to expend too many LP iterations (and hence time) to obtain the optimal solution, but would require more branching operations to do so and hence more number of nodes. This is because at the very beginning of the BnB tree, the pseudocost branching scheme has no history to use for guiding its choice on branching. Since the branching decisions near the top of the BnB tree are the most crucial, the absence of early history information would lead to more node explorations. If the optimal solution is not the main objective and obtaining solutions with low duality gap is desired in a given time, it is expected that strong branching would return a large duality gap since the rate of improvement is expected to be slow for the given solution time due to the high number of LP iterations required. A hybrid branching technique, which combines both schemes, aims to get the best of each and reduce as much as possible the cons each has. It is achieved by implementing strong branching in the upper part of BnB tree up to a certain depth d. For nodes that are deeper than d in the tree, pseudocost branching is applied.

4.3.6 Reliability Branching

The branching decision based on pseudocosts in either pure pseudo-cost branching or hybrid strong/pseudocost branching are based on uninitialized values that negatively affect the selection of branching variables. Reliability branching uses strong branching for variables with uninitialized pseudocosts (defined in Section 4.3.3) and hence is more dynamic than *hybrid strong/pseudocost branching* which uses *strong branching* for a fixed depth in the BnB tree. Furthermore, to use the pseudocosts for branching, reliability branching requires that the history for the branching variable be collected for at least η_{rel} problems, where η_{rel} is the reliability parameter. Hence, if $min\left\{\eta_j^0, \eta_j^1\right\} \leq \eta_{rel}$ the variable $\ddot{x}_j$ is called unreliable. Moreover, the work expended in strong branching can be reduced using a small subset of branching variable candidates $\overline{F} \subset F_{cand}$ as well as performing only a few simplex iterations for each candidate in $\overline{F}$ to estimate the changes in the dual bound. The dual bound is the value of the objective function of $\mathcal{Q}_{relax}$. Since the change in the objective function value is greatest in the first few simplex iterations compared to later iterations, the estimate for the dual bound is expected to be close to the actual value.

The η_{rel} dynamically changes to restrict the number of strong branching simplex iterations for a given node $\mathcal{Q}$ to [63] :

$$\hat{\gamma}_{SB}^{max} = c_{sbiterquot}\hat{\gamma}_{LP} + \hat{\gamma}_{SB}^{root} + \hat{\gamma}_{fixed}, \tag{4.8}$$

where

- $\hat{\gamma}_{SB}^{max}$ is the number of simplex iterations for for the strong branching done in $\mathcal{Q}$
- $\hat{\gamma}_{LP}$ is the number of regular simplex iterations
- $c_{sbiterquot}$ is maximal fraction of strong branching LP iterations compared to node relaxation LP iterations,
- $\hat{\gamma}_{fixed}$ is a fixed number that can be pre-set.

If the number of strong branching LP iterations $\hat{\gamma}_{SB}$ exceeds $\hat{\gamma}_{SB}^{max}$, then η_{rel} is set to zero and pseudocost branching is used. If $\hat{\gamma}_{SB} \in [c_{sbiterquot}\hat{\gamma}_{SB}^{max}, \hat{\gamma}_{SB}^{max}]$, η_{rel} decreases from η_{rel}^{max} to η_{rel}^{min} linearly. If $\hat{\gamma}_{SB} < c_{sbiterquot}\gamma_{LP}$, then η_{rel} increases in proportion to the quotient $\frac{\hat{\gamma}_{LP}}{\hat{\gamma}_{SB}^{max}}$.

4.3.7 Inference Branching

This technique exploits domain propagation of the branching variables. Its main idea is that it selects the variable whose domain tightening (variable fixation in case of binary variables) produces the most domain reductions in other variables. The impact of a variable on domain deductions is obtained from history information, like pseudo-cost branching, that measures the average inferred domain deductions $\ddot{\Phi}_j^1$ and $\ddot{\Phi}_j^0$ given by [62]:

$$\ddot{\Phi}_j^1 = \frac{\ddot{\phi}_j^1}{\ddot{\nu}_j^1} \qquad and \qquad \ddot{\Phi}_j^0 = \frac{\ddot{\phi}_j^0}{\ddot{\nu}_j^0}, \tag{4.9}$$

where

- $\ddot{\phi}_j^1$ and $\ddot{\phi}_j^0$ are the total deductions by setting the binary variable $\ddot{x}_j$ to 1 or 0, respectively,
- $\ddot{\nu}_j^1$ and $\ddot{\nu}_j^0$ are the numbers of corresponding subproblems for which domain propagation has been applied.

For uninitialized binary variables, clique and implication tables are used to calculate the inference values [62].

4.3.8 Cloud Branching

All branching strategies described above deal with only one optimal fractional solution for $\mathcal{Q}_{relax}$. While LP relaxations are known to be largely degenerate, multiple equivalent optimal solutions are the rule rather than the exception. Therefore, considering only one optimal solution yields high possibilities of taking arbitrary, or inefficient branching decisions. Cloud branching exploits the knowledge of a cloud of multiple alternative optimal solutions of the given LP relaxation using dual degeneracy in a mixed integer program [64]. For a given cloud $\mathfrak{C} = \left\{ \ddot{x}^1, ... \ddot{x}^k \right\}$ of optimal solutions of the LP relaxation, the initial set of branching variable candidates $F(\mathfrak{C})$ contains all the variables that are fractional in at least one solution of the cloud $\mathfrak{C}$. The cloud of solutions is generated in the context of strong branching, which solves the LPs that would result from branching on all candidates.

The first step in the cloud branching strategy is to generate a cloud of alternative optimal solutions for the LP relaxation $\mathcal{Q}_{relax}$ of a node $\mathcal{Q}$. This is done by restricting search for the basic feasible variables to the optimal hyperplane of the polyhedron. To implement this type of search, the variables of a given optimal solution, whose reduced costs are non-zero need to be fixed in the search procedure. To move from one basis to another on the optimal hyperplane, an auxiliary objective function is needed. The one used in our numerical experiments is a feasibility like pump objective function that is implemented in the SCIP solver and was proposed in [64] whose coefficients for the binary variables $j \in \mathcal{B}$ are given as:

$$c_j = \begin{cases} -1 & if \ 0 < \ddot{x}_j^* < 0.5 \\ 1 & if \ 0.5 \le \ddot{x}_j^* < 1. \end{cases} \tag{4.10}$$

Using iterations of the primal simplex algorithm on the resulting auxiliary LP $\mathcal{Q}_{Aux}$, an alternative optimum basis to the LP relaxation of the BnB node can be obtained that has the closest hamming distance to the nearest integral point.

After obtaining a cloud $\mathfrak{C}$, the cloud interval for a variable $\ddot{x}_j \in F(\mathfrak{C})$ is given by $\left[l_j^{\mathfrak{C}}, u_j^{\mathfrak{C}} \right]$, where:

$$l_j^{\mathfrak{C}} = min \left\{ \ddot{x}_j^i | \ddot{\mathbf{x}}^i \in C \right\}, \tag{4.11a}$$

$$u_j^{\mathfrak{C}} = max \left\{ \ddot{x}_j^i | \ddot{\mathbf{x}}^i \in C \right\}. \tag{4.11b}$$

Accordingly, the set $F(\mathfrak{C})$ is partitioned into three which are:

$$F_2 = \left\{ j \in F(\mathfrak{C}) \, | 0 < l_j^{\mathfrak{C}} \wedge u_j^{\mathfrak{C}} < 1 \right\}, \tag{4.12a}$$

$$F_0 = \left\{ j \in F(\mathfrak{C}) \, | l_j^{\mathfrak{C}} = 0 \wedge u_j^{\mathfrak{C}} = 1 \right\}, \tag{4.12b}$$

$$F_1 = F(\mathfrak{C}) \setminus (F_2 \cup F_0), \tag{4.12c}$$

which shows that for binary variables, the only type of discrete variables in $\mathcal{HAP}_{MBQCP}^{Eff}$, F_2 contains the fractional variables of all the solutions in the cloud $\mathfrak{C}$.

Branching on the variables in F_0 guarantees that the dual bound in both branching directions will not improve. Those in F_1 are guaranteed not to improve the bound in only one direction but hopefully will improve in the other direction. The candidates in F_2 are expected to to improve the dual bound in both directions. The cloud purpose is to filter out as many LPs so that strong branching only needs to solve a small subset of those. As long as there are any candidates existing in the set F_2, the other two sets are ignored and only the LPs for the candidates in F_2 are solved.

4.4 Computational Experiments and Results

This section discusses the experiments conducted for $\mathcal{HAP}^{Eff}_{MBQCP}$ and presents the numerical results obtained for the BnB algorithm procedures given in Sections 4.1 and 4.3 to evaluate their performances. Two of the conducted experiments are discussed in this section. The first experiment (Section 4.4.1) compares the performance of activating-versus-deactivating the reformulation linearizion technique, at the presolving phase, for the quadratic constraint set $\overline{C8}$ in $\mathcal{HAP}^{Eff}_{MBQCP}$ which was explained in Section 4.1. The second experiment set (Section 4.4.2) compares the performance of the different branching techniques explained in Section 4.3. The performance for each set of experiments is measured using the following criteria:

1. the duality gap,
2. number of LP iterations expended,
3. number of nodes in the search tree and
4. average number of LP iterations per node.

The experiments were performed in Matlab, for which the open source optimization toolbox OPTI (version 2.16) [65] provided the interface with the SCIP 3.2.0 solver [59]. SCIP 3.2.0 is the solver used in all the experiments conducted for $\mathcal{HAP}^{Eff}_{MBQCP}$. The experiments were performed on a machine with a 6 core 3.5 GHz Intel Xeon processor. Using the parallel processing toolbox in Matlab [66], we were able to conduct different experiments in parallel. For example, to conduct experiments on different branching strategies, each CPU

Table 4.1 SCIP solver settings for all experiment sets conducted

Parameter	Value
Solving time limit	10 Minutes
LP iteration limit per node	10^5 iterations
BnB node limit	10^7 nodes
Feasibility tolerance	1^{-12}
Integrality tolerance	1^{-7}

core performed the experiment of a specific branching strategy for the same set of problem instances in parallel to the other cores. The generic SCIP solver settings used in all the experiment sets performed are given in Table 4.1.

One hundred instances were solved for each experiment. Each instance has a size of 527 variables and 4261 constraints out of which 107 variables are binary and 2844 constraints are quadratic. To obtain the channel gain $g_{i,k,c,t}$ values, a simulation was conducted using the parameters in Table 4.2, equation (3.4) in this book and (13), (14) and (15) from our earlier work in [4]. The channel consists of a free space pathloss propagation model, Ricean fading as suggested in [5], the rain attenuation model in [58], the HAP aperture antenna model in [24] and parabolic reflector user antenna models. In the simulation, the user positions change during every iteration according to a uniform probability distribution about the HAP footprint centers. The degree of overlap between antenna footprints is defined as the ratio between the overlap distance $d_{overlap}$, illustrated in Figure 4.2, and the HAP antenna footprint radius $r_{footprint}$. Figure 3.1 in section 3.1 illustrates the overlapped HAP antenna footprints in our experiments.

To evaluate the average performance of all the instances for each experiment, arithmetic, geometric and shifted geometric means were used. For the shifted geometric mean, the shifting parameters values used are:

1. 50, for the dual gap,
2. 100, for the number of BnB nodes,
3. 1000, for the number of LP iterations.

Table 4.2 Simulation parameters for multicasting in a single HAP based network

Parameter	Value
Number of multicasting sessions (M)	2
Number of antennas on board (S)	7
Number of users in the service area (K)	10
Number of available subchannels (C)	3
Number of available time slots (T)	2
HAP height	20 Km
Degree of antenna beam footprint overlap	105 %
HAP antenna footprint radius ($r_{footprint}$)	500 meters
HAP antenna side lobe level	-40 dB
SINR threshold ($\gamma_{m,i}^{th}$)	35
Noise power spectral density (N_o)	-173 dBm/Hz
Maximum rate requirements (R_m^{max})	20 Mbps
Minimum rate requirements (R_m^{min})	10 Mbps
Carrier frequency	2.1 GHz
Total HAP power (P_{PF}^{Total})	1 Watt
OFDMA frame length	20 ms
Total bandwidth	15 MHz
Rice factor (dB)	20 dB
Rain attenuation factor (χ)	3 dB/Km
Set of values for the user-session priority levels ($\rho_{m,k}$)	$\rho_{m,k} \in \{1, 2, 3, 4, 5\}$
The binary constants indicating the admission request of user k for session m ($\lambda_{m,k}$)	$\lambda_{m,k} = 1 \ \forall \ m, k$
User antenna diameter (D_{user}^{Ant})	$0.75m$

The shifted geometric mean of a sample $w_1, w_2, ..., w_k$ is given by [62]:

$$\psi_s = \left(\prod_{j=1}^{k} max\, \{w_j + s, 1\} \right)^{1/k} - s, \qquad (4.13)$$

where s is the shifting parameter. For geometric mean, $s = 0$. In the comments made on the results in the following subsections, we use the shifted geometric means for comparison except for the average number of LP iterations per node which uses only arithmetic means.

The duality gap is calculated in all the experiments, in percentage, using the formula:

$$\varrho = \frac{|\ddot{c}_{dual} - \ddot{c}_{primal}|}{min\left(|\ddot{c}_{dual}|, |\ddot{c}_{primal}|\right)}. \qquad (4.14)$$

4.4.1 Reformulation Linearization at the Presolving Phase

In this sub-section, the experimental procedures and results for the reformulation linearization technique explained in Section 4.1 are provided. The reformulation technique is invoked at the presolving phase, and hence the experiments illustrate the performance of activating-versus-deactivating the linearization for the number of presolving rounds 1, 5, 25, 50 and 100. The following settings were considered for the reformulation linearization experiments:

1. node selection scheme is *best first search* with a maximum plunging depth in the BnB tree of 2 [59],
2. *most infeasible branching* scheme was used and
3. the only heuristic used was the *Undercover* heuristic [67].

In Figures 4.3, 4.4, 4.5 and 4.6 the duality gap, the number of LP iterations, the number of BnB nodes and the average number of LP iterations per node are illustrated respectively, for the experiments. In those figures, RndNum$i$:'.', indicates the number of presolving rounds is i for either: 'A', active reformulation linearization or 'D', deactivated reformulation linearization.

We can see that for a single presolving round, the dual gap is almost the same in both 'A' and 'D'. The number of BnB nodes is slightly lower by around 11% for 'A' compared to 'D'. The number of LP

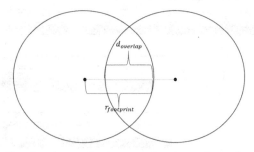

Figure 4.2 Overlap degree for two overlapping antenna beam footprints.

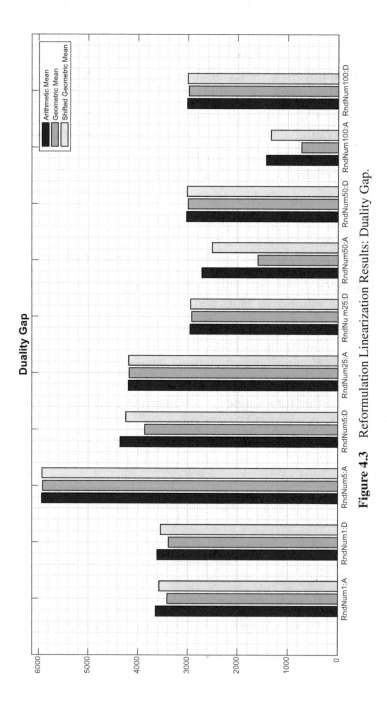

Figure 4.3 Reformulation Linearization Results: Duality Gap.

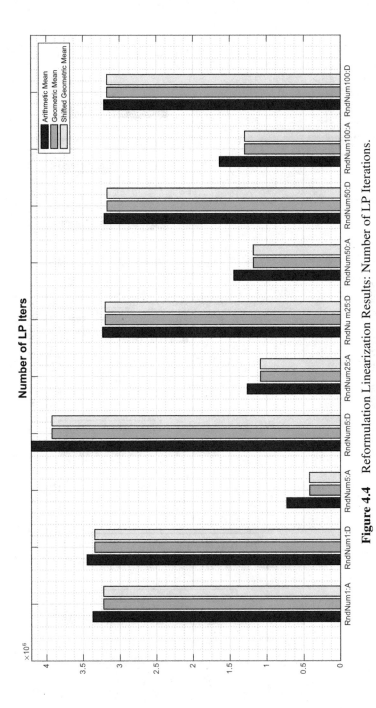

Figure 4.4 Reformulation Linearization Results: Number of LP Iterations.

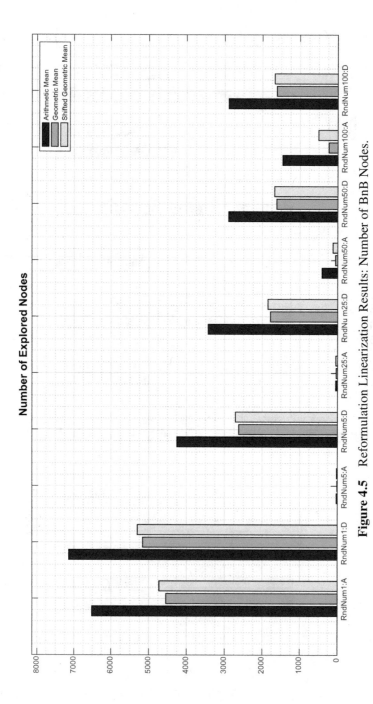

Figure 4.5   Reformulation Linearization Results: Number of BnB Nodes.

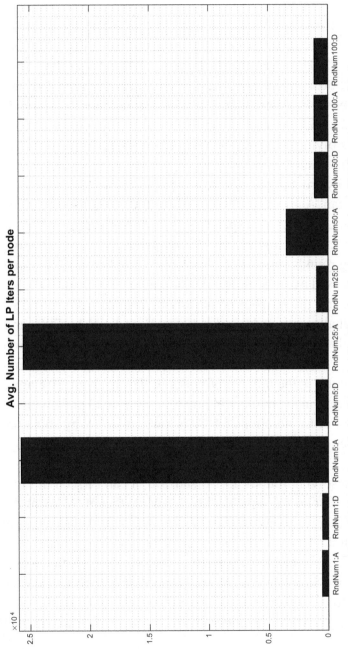

Figure 4.6 Reformulation Linearization Results: Average number of LP Iterations per Node.

iterations and the average number of LP iterations per node are almost equal for 'A' and 'D'.

A small increase in the number of presolving rounds from 1 to 5 yields an increase in the dual gap for both cases 'A' and 'D' as shown in Figure 4.3. However, the dual gap for 'D' is lower than that of 'A' by around 30% at the expense of a much larger number of nodes in comparison to 'A' (more than 2700 %). The number of explored nodes is lower for both 'A' and 'D' for five presolving rounds as compared to one presolving round as shown in Figure 4.5. For 'A', it is lower by around 99 % and 'D' is lower by 50 %. The number of LP iterations decreases by around 85% for 'A' and increases by 15% for 'D' making it higher than 'A' by almost 800%. According to Figure 4.6, the number of average LP iterations expended per node for five presolving rounds is around 2500 % higher for 'A' compared to 'D'. Comparing five presolving rounds versus one, the average number of LP iterations per node increased by 5100 % for 'A' but only by 100% for 'D'.

Increasing the number of presolving rounds from 5 to 25 shows that the duality gap reduces for both 'A' and 'D' by 29% and 31% respectively (almost same reduction), while 'D' still has a lower dual gap by 30% compared to 'A'. The reduction in dual gap is accompanied by a reduction in the number of nodes in 'D' by about 31% and a very small increase in the number of nodes in 'A'. For 25 presolving rounds, the number of LP iterations for 'A' is lower than 'D' by about 65% but the number of average LP iterations per node is much higher by about 2450%.

For presolving rounds 25, 50 and 100, it can be seen in Figure 4.3 that 'D' maintains the same duality gap while that of 'A' keeps decreasing. For 100 presolving rounds, we can see that 'A' has a duality gap lower than 'D' by 60 %. The number of BnB nodes gradually decreases slightly for 'D' when increasing the presolving rounds in the range 25, 50 and 100 while that for 'A' keeps increasing such that the number of nodes for 100 rounds increases by about 900 %. However, at 100 presolving rounds, the number of nodes for 'A' is lower than 'D' by about 70%. Figure 4.4 shows that the number of LP iterations for presolving rounds 25, 50 and 100 remains approximately the same for 'D' but increases slightly for 'A'. For presolving rounds 5, 25, 50 and 100, it can be seen from Figure 4.6 that the average number of LP iterations per node is greatly reduced as the number of rounds for 'A' and becomes equivalent to 'D' whose average LP iterations per node remains almost the same for rounds 5, 25, 50 and 100.

From the results, we can hence conclude that it is beneficial to use the reformulation linearizion technique for constraint set $\overline{C8}$ with a high number of presolving rounds (around 100).

4.4.2 Branching Schemes

In this sub-section, the experimental procedures and results for the branching techniques discussed in Section 4.3 are provided. Strong Branching is not considered by itself in the experiments due to its expected high computational effort and time. However, as explained in Section 4.3, it is a component of hybrid strong/psedocost, reliability and cloud branching, where its effect will be seen in those branching schemes. The following settings are considered for the experiments conducted for the branching schemes:

1. separating cuts are deactivated,
2. node selection scheme is best first search with a maximum plunging depth in the BnB tree of 2 [59],
3. up to one round of presolving before starting the BnB algorithm,
4. for hybrid strong/pseudocost branching, maximum strong branching depths of $\tilde{d}_{strong} = 1$ and $\tilde{d}_{strong} = 2$ are tried in two different experiments, and
5. for reliability branching, the following settings are considered:
   - The maximum value for the reliability threshold is $\eta_{rel}^{max} = 5$,
   - the minimum value for the reliability threshold is $\eta_{rel}^{min} = 1$,
   - $\hat{\gamma}_{fixed} = 0$ in Equation (4.8),

- maximum size of the set of strong branching candidates, $\overline{F} = 15$,
- maximum number of strong branching simplex iterations per branching variable is $\hat{\gamma}_{sbiterbrancand} = 100$,
- the ratio $c_{sbiterquot}$ in Equation (4.8) is set to $c_{sbiterquot} = 0.05$ and $c_{sbiterquot} = 0.2$ for two different experiments.

Figures 4.7, 4.8, 4.9 and 4.10 show the duality gap, number of LP iterations, number of BnB nodes and the average number of LP iterations per node for the different branching schemes. In those figures, HybDepth1 and HybDepth2 are the hybrid strong/pseudocost branching with strong branching invoked up to maximum depths of 1 and 2, respectively. Furthermore, $Relratio = 0.05$ and $Relratio = 0.2$ refer to *reliability branching* with $c_{sbiterquot} = 0.05$ and $c_{sbiterquot} = 0.2$. It can be seen that random branching has the highest duality gap, which is expected since the selection of branching candidates does not take into account the direction of change of the dual bound. The lowest duality gap was achieved (almost equally) by inference branching, pseudocost branching, hybrid strong/pseudocost branching ($\tilde{d}_{strong} = 1$) and surprisingly most infeasible branching. The second lowest are the cloud branching and reliability branching with $c_{sbiterquot} = 0.05$ equally both having a higher duality gap than the lowest four by about 18 %. Finally, the second highest duality gap is obtained by *reliability branching* with $c_{sbiterquot} = 0.2$ with a duality gap higher than the lowest four by 50%.

Comparing pseudocost branching versus hybrid strong/pseudocost branching with $\tilde{d}_{strong} = 1$, it can be seen that they almost perform equally in terms of duality gap, number of expended LP iterations, number of nodes and the average LP iterations per node. Increasing the depth for strong branching to $\tilde{d}_{strong} = 2$ in hybrid strong/pseudocost branching leads to an increase in the duality gap by 32%. This is because when more strong branching is involved, a slightly greater number of LP iterations per node are expended as shown in Figure 4.10, meaning that in a given time limit, fewer nodes are explored as shown in Figure 4.9. When fewer nodes are explored

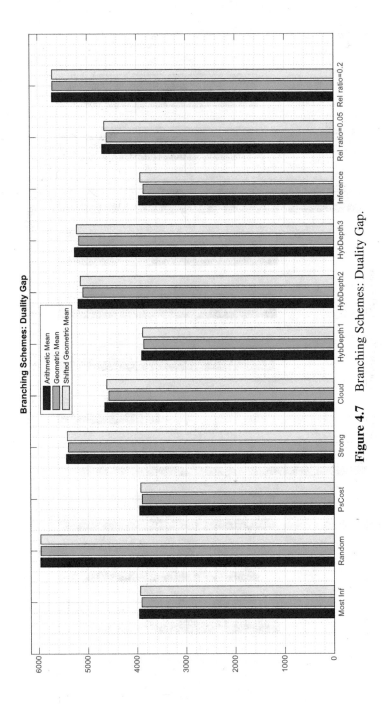

Figure 4.7 Branching Schemes: Duality Gap.

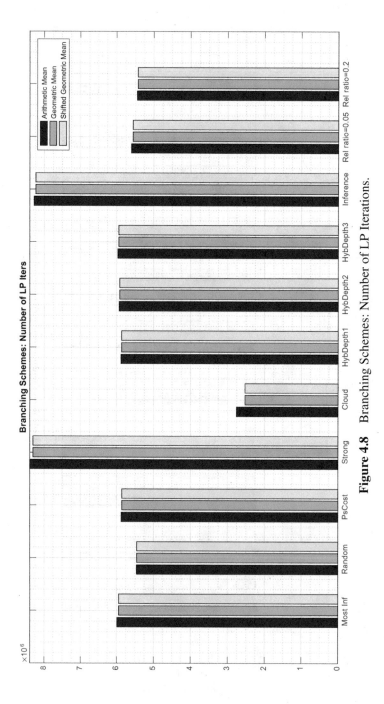

Figure 4.8   Branching Schemes: Number of LP Iterations.

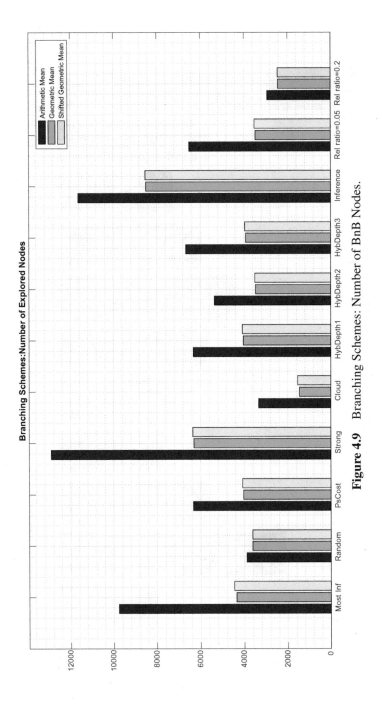

Figure 4.9   Branching Schemes: Number of BnB Nodes.

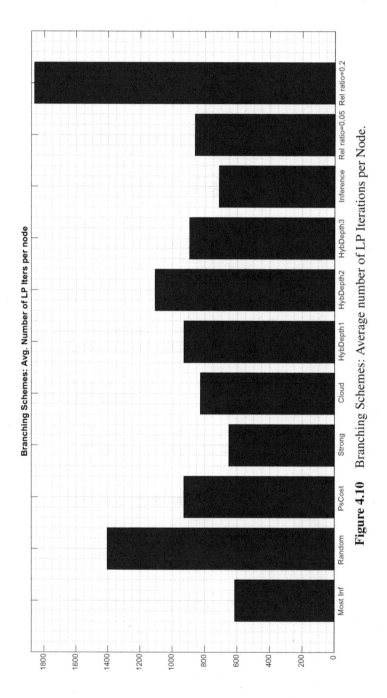

Figure 4.10   Branching Schemes: Average number of LP Iterations per Node.

for a given time limit, the overall dual bound improvement could be lower, even though the improvement per node can be higher for strong branching. The same reasoning applies for reliability branching in the two experiments in which $c_{sbiterquot} = 0.05$ and $c_{sbiterquot} = 0.2$.

Among the four branching schemes that give the lowest duality gaps, inference branching needs the largest number of nodes and LP iterations as shown in Figures 4.9 and 4.8. Cloud branching requires the lowest number of nodes and LP iterations among all the branching schemes but has the second lowest duality gap. It requires 64 % less number of nodes and 58 % less number of LP iterations compared to the most infeasible branching. Although cloud branching is based on strong branching, the cloud reduces out many LPs so that strong branching solves a small subset of those. It hence gives a better duality gap than HybDepth2 and reliability branching at $c_{sbiterquot} = 0.2$ requiring lower number of BnB nodes and LP iterations. It also gives an equally good duality gap for lower number of BnB nodes and LP iterations compared to *reliability* branching with $c_{sbiterquot} = 0.05$.

According to the observations and results' analysis of Figures 4.7, 4.8, 4.9 and 4.10, *cloud branching* seems to have a good trade-off balance of all the criteria of interest.

4.5 Chapter Conclusion

In this chapter, we presented parts of the branch and bound-based solution framework proposed for solving $\mathcal{HAP}^{Eff}_{MBQCP}$. We considered two components of the solution framework, a presolving reformulation linearization for a particular set of quadratic constraints in the formulation as well as various branching schemes. For the presolving reformulation linearization, the aim was to evaluate the tradeoffs and find whether it would make the solution procedure more efficient. According to the results obtained from the experiments that we conducted, we found that reformulation linearization is recommended while applying at least 100 presolving rounds as this gave the lowest duality gap with lower number of LP iterations and nodes as compared to not applying the reformulation linearization technique. For the branching schemes, we showed that cloud branching has a good trade-off between the duality gap, number of LP iterations and number of BnB nodes when compared with the other branching schemes.

5

Proposed Solution Method: Cutting Planes, Domain Propagation and Primal Heuristics

In Chapter 4, we focused on two aspects of the solution framework, one was applying presolving reformulation linearization for a particular specific class of quadratic constraints that appears in the formulation of the problem. The other aspect was the use of different types of branching techniques. This chapter focuses on cutting planes, domain propagation and different heuristics embedded in the BnB solution framework. These are explained in the context of the formulated problem and their performances are presented for the computational experiments that we conducted for $\mathcal{HAP}^{Eff}_{MBQCP}$. Figure 5.1 shows the two components that were considered in Chapter 4 represented by the green colored boxes, while the three components considered in this chapter are represented by the three blue boxes.

5.1 Cutting Planes and Cut Separation Process

In Chapter 4, we mentioned that the relaxation procedure is one that can greatly affects the algorithm's computational effort, speed and BnB tree size. For the quadratic non-convex constraints in $\mathcal{HAP}^{Eff}_{MBQCP}$, linear relaxation is achieved using linear outer approximation with McCormick underestimators for all the bilinear terms of the quadratic constraints. The relaxation can be strengthened using cutting planes that separate sets of solutions from a BnB node relaxation $\mathcal{Q}_{relax}$ including its optimal solution $\ddot{x}^{opt}_{relax}$ without removing any of the feasible solutions for node $\mathcal{Q}$. Adding cutting planes to the LP

89

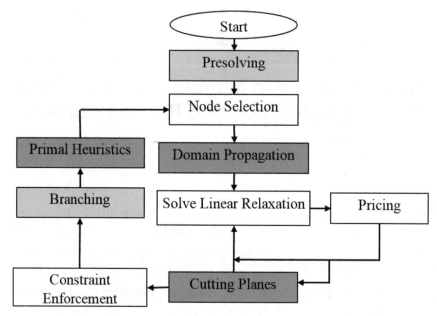

Figure 5.1 Flowchart illustrating the interrelation between the SCIP solver components used for solving the HAP multicasting AC-RRA problem.

relaxation of a subproblem simply changes the simplex tableau by adding new rows that represent the cuts added.

Since the addition of new rows to the simplex tableau does not affect the LP dual feasibility of the solution $\ddot{\mathbf{x}}_{relax}^{opt}$, it could be used as a warm up start for the dual simplex algorithm to solve the strengthened relaxation $\mathcal{Q}_{relax}^{Strengthened}$ of the node $\mathcal{Q}$. This could be repeated several times until a maximum number of LP iterations is achieved, a maximum LP time limit is hit, a maximum number rounds of cut separation or a maximum number cuts is achieved. This procedure utilizes a very important advantage of LP solvers, which is warm up starts. Warm-up starts mean that instead of starting all over again after adding cutting planes, an optimal solution of a node relaxation can be used to find the new optimal solution of $\mathcal{Q}_{relax}$. This definitely could save a lot unnecessary iterations and helps improve the dual bound $\ddot{c}_{dual}$ in the BnB algorithm leading to the pruning of large sections of the BnB tree.

Figure 5.2 illustrates what the insertion of a cutting plane into an LP relaxation does. This illustration assumes two binary variables whose axes are horizontal and vertical besides one continuous variable whose axis is perpendicular to the page. Hence, what we see in the figure is a cross-section of the feasible region where the black dots are actually parallel lines emanating from the page. Note that in the figure, the bilinear terms of the non-convex quadratic constraints in $\mathcal{HAP}_{MBQCP}^{Eff}$ are considered to be linearly relaxed by McCormick under-estimators. The blue region represents the convex hull of the discrete feasible solutions to a small problem instance. A convex hull $clcnv\,(.)$ is the smallest convex set that includes the set of all feasible solutions. The union of the green area and the blue represents the relaxation of the subproblem $\mathcal{Q}_{relax}$ and its optimal solution $\ddot{x}_{relax}^{opt}$.

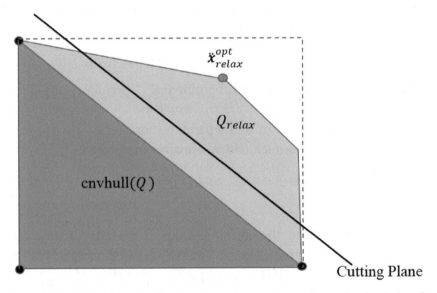

Figure 5.2 Illustration of a cutting plane that separates the optimal solution for $\mathcal{Q}_{relax}$ (represented by the red dot) from the convex hull of $\mathcal{Q}$ (represented by the blue triangle)

The purpose of the cutting plane is to tighten the relaxation by cutting into the feasible region of $\ddot{\mathbf{x}}_{relax}^{opt}$ as deep as possible without passing into the $clcnv\,(\mathcal{Q})$ which contains the feasible solutions of $\mathcal{Q}$.

Cutting plane separation is performed in rounds, where in each round, many different cutting planes are generated to cut off $\ddot{\mathbf{x}}_{relax}^{opt}$. All the cuts are stored in a *separating storage* from which only a subset of those are selected for cutoff. Just as in [62], the cuts are selected based upon the following:

- the *efficacy* $\overline{eff}_c$ of the cut, which is the Euclidean distance between the cut and $\ddot{\mathbf{x}}_{relax}^{opt}$,
- the *orthogonality* $\overline{orth}_c$ of the cuts with respect to each other and
- the *parallelism* $\overline{par}_c$ of the cuts with respect to the objective function, which is linear.

A score $s_{cut}\left(\overline{eff}_c, \overline{orth}_c, \overline{par}_c\right)$ based upon the weighted sum of these criteria is given by [62]:

$$s_{cut}\left(\overline{eff}_c, \overline{orth}_c, \overline{par}_c\right) = w_{eff}\overline{eff}_c + w_{orth}\overline{orth}_c + w_{par}\overline{par}_c. \quad (5.1)$$

If the solution of a relaxed subproblem $\ddot{\mathbf{x}}_{relax}^{opt}$ violates any of the quadratic constraints in $\mathcal{HAP}_{MBQCP}^{Eff}$, which are all non-convex, each bilinear term gets underestimated separately. For a term with a positive coefficient a, the McCormick underestimators are given as [57]:

$$a\ddot{x}_j\ddot{x}_k \geq a\ddot{x}_j^L\ddot{x}_k + a\ddot{x}_k^L\ddot{x}_j - a\ddot{x}_j^L\ddot{x}_k^L, \quad (5.2a)$$

$$a\ddot{x}_j\ddot{x}_k \geq a\ddot{x}_j^U\ddot{x}_k + a\ddot{x}_k^U\ddot{x}_j - a\ddot{x}_j^U\ddot{x}_k^U, \quad (5.2b)$$

where $\ddot{x}_j^L$ and $\ddot{x}_j^U$ are the lower and upper bounds of any variable $\ddot{x}_j$ respectively and $\ddot{x}_j$ is any of the decision variables in the problem $\mathcal{HAP}_{MBQCP}^{Eff}$. If $\left(\ddot{x}_j^U - \ddot{x}_j^L\right)\ddot{x}_{relax_k}^{opt} + \left(\ddot{x}_k^U - \ddot{x}_k^L\right)\ddot{x}_{relax_j}^{opt} \leq \ddot{x}_j^U\ddot{x}_k^U - \ddot{x}_j^L\ddot{x}_k^L$ inequality (5.2a) is used, otherwise inequality (5.2b) is used. If the bilinear term coefficient a is negative, the McCormick underestimators are:

$$a\ddot{x}_j\ddot{x}_k \geq a\ddot{x}_j^U\ddot{x}_k + a\ddot{x}_k^L\ddot{x}_j - a\ddot{x}_j^U\ddot{x}_k^L, \quad (5.3a)$$

$$a\ddot{x}_j\ddot{x}_k \geq a\ddot{x}_j^L\ddot{x}_k + a\ddot{x}_k^U\ddot{x}_j - a\ddot{x}_j^L\ddot{x}_k^U, \quad (5.3b)$$

If $\left(\ddot{x}_j^U - \ddot{x}_j^L\right)\ddot{x}_{relax_k}^{opt} - \left(\ddot{x}_k^U - \ddot{x}_k^L\right)\ddot{x}_{relax_j}^{opt} \leq \ddot{x}_j^U \ddot{x}_k^L - \ddot{x}_j^L \ddot{x}_k^U$ inequality 5.3a is used, otherwise inequality 5.3b is used.

Besides the McCormick separators, we use implied cuts, clique cuts [62] as well as the generic mixed integer Gomory cuts [68] for $\mathcal{HAP}_{MBQCP}^{Eff}$. For implied bound cuts, the separator inspects the implication graph to extract cuts that are violated by $\ddot{\mathbf{x}}_{relax}^{opt}$. These implications can only be violated if a binary variable has a fractional value in $\ddot{\mathbf{x}}_{relax}^{opt}$. The implication graph is a directed graph whose purpose is to store the strongest implications between variables in $\mathcal{HAP}_{MBQCP}^{Eff}$. An implication is simply a derivation of the values of other variables if a variable $\ddot{x}_j$ is set to a specific value (or range of values). For example, if we consider constraint set $\overline{C5}$ in formulation $\mathcal{HAP}_{MBQCP}^{Eff}$, we can see that if $y_{m,i,c,t} = 0$ this implies $p_{m,i,c,t} = 0$. A second example is $\overline{C9}$ which implies $\theta_m = 0$ when all the corresponding $y_{m,i,c,t}$ in the constraint are equal to zero. Many similar implications are deduced in the presolving stage from which the implication graph is constructed.

For clique cuts, a clique graph that stores sets of binary variables where for each set, only one (complemented) binary variable can be set to 1 (zero) while the rest of the (complemented) variables in the set should be set to zero (1). A clique inequality has the form

$$\sum_{Q_{binary}} \ddot{x}_j \leq 1, \tag{5.4}$$

where $Q_{binary} \subset \mathcal{B} \cup \overline{\mathcal{B}}$ is a subset of binary variables and their complements. In $\mathcal{HAP}_{MBQCP}^{Eff}$, some cliques could be obtained directly from the set packing constraint set $\overline{C2}$ which are cliques by themselves. Presolving could further deduce more cliques while simplifying other constraints to the point that they could be upgraded to set packing constraints or using probing [69]. The LP values of binary variables are used as weights for the nodes of the clique graph for the separation of violated cliques using *T-Clique* [70] algorithm in SCIP for solving $\mathcal{HAP}_{MBQCP}^{Eff}$.

5.2 Domain Propagation

After a node in the BnB tree gets selected for processing and primal heuristic methods have been called, domain propagation methods are called for different constraint types to tighten the variable's local domains, or if possible fix their values. This is performed iteratively in rounds until no more domain reductions are possible or until a maximum number of propagation rounds is hit. The inferred domain deductions can yield stronger linear underestimators in the separation process (for quadratic constraints), they can cutoff nodes due to the infeasibility of a constraint and can result in further domain deductions on other constraints. There are different propagation methods in SCIP that could be tailored for the different types of constraints in $\mathcal{HAP}^{Eff}_{MBQCP}$ as explained in the following sections.

5.2.1 Domain Propagation Schemes for Quadratic Constraints

For the quadratic constraints, we use an interval-arithmetic based method in SCIP [71]. Forward and backward propagations are invoked for a quadratic constraint. Forward propagation is a step that is done to reduce a quadratic constraint interval $[l_c, u_c]$ which, using simple interval arithmetic, yields the new reduced interval for the constraint $[l_c^{new}, u_c^{new}]$ with respect to its variable domains. If the intersection between the new constraint interval and the old constraint is empty, i.e. $[l_c, u_c] \cap [l_c^{new}, u_c^{new}] = \emptyset$, the BnB node $\mathcal{Q}$ could be pruned, otherwise the constraint domain could be reduced to $[l_c, u_c] \cap [l_c^{new}, u_c^{new}]$.

Backward constraint propagation infers domain deductions on the variables in a quadratic constraint given the constraint interval $[l_c, u_c]$. This is achieved by solving the quadratic interval equation for the quadratic constraint and its interval whose solutions for all its bilinear (or quadratic) variables are the intervals $[l^{\ddot{x}_j}, u^{\ddot{x}_j}]$ [71]. If the intersection between the old variable bounds and the new inferred variable intervals is empty, i.e. $[l^{\ddot{x}_j}, u^{\ddot{x}_j}] \cap [\ddot{x}_j^L, \ddot{x}_j^U] = \emptyset$, then the BnB node $\mathcal{Q}$ can be pruned, otherwise the new variable bounds inferred are $\left[l^{\ddot{x}_j}, u^{\ddot{x}_j}\right] \cap \left[\ddot{x}_j^L, \ddot{x}_j^U\right]$. More details on the forward and backwards quadratic constraints propagation are in [57].

5.2.2 Domain Propagation Schemes for Linear Constraints

In the formulation $\mathcal{HAP}^{Eff}_{MBQCP}$ constraint sets $\overline{C1}-\overline{C7}$ and $\overline{C9}-\overline{C10}$ are all linear sets. Constraint sets $\overline{C2}$ and $\overline{C5}$ are explicitly set packing and variable bound constraints, respectively. Furthermore, during the presolving phase, many linear constraints in a given instance of $\mathcal{HAP}^{Eff}_{MBQCP}$ get reduced to set packing, variable bound and set covering constraints. For example, one of the instances passed to SCIP in the experiments conducted had 256 linear constraints and 2844 quadratic constraints. After the presolving phase, they were reduced to 2004 quadratic constraints at the expense of 5882 general linear constraints, 146 set packing and covering constraints and 2604 variable bound constraints. The general linear constraints propagators that we use for $\mathcal{HAP}^{Eff}_{MBQCP}$ use the concepts of *activity bounds* and *activity bound residuals*, for which we refer the reader to the PhD. thesis of T. Achterberg [62] for a detailed explanation of the procedure. In the following, we explain more specific linear types of constraint and objective function propagators.

1. **Set Packing Constraints:** Any set packing constraint (including $\overline{C2}$) taking the form of Equation (5.4) and has only one possibility of domain propagation, that is:

$$\ddot{x}_k = 1 \qquad \rightarrow \forall j \in Q_{binary} \setminus \{k\} : \ddot{x}_j = 0 \qquad (5.5)$$

For this type of constraint, if two or more variables in the set Q_{binary} are fixed to one, the subproblem Q is then infeasible and can be pruned. If exactly one variable in the set Q_{binary} is fixed to one, then the rest of the variables in the set must be zero. In this case, the constraint gets deleted from the subproblem. This is achieved by keeping a track and updating a counter of the current number of variables fixed to 1 in the constraints in $\overline{C2}$.

2. **Set Covering Constraints:** During the course of the BnB algorithm, at each time a pair of subproblems is created, a binary variable is either fixed to one or to zero. If any of the binary variables θ_m in $\mathcal{HAP}_{MBQCP}^{Eff}$ is fixed to 1 in any node in the tree, the corresponding constraint in the constraint set $\overline{C3}$ becomes a set-covering constraint if none of the variables $y_{m,i,c,t}$ in that constraint was fixed to one. If any of the variables $y_{m,i,c,t}$ in $\overline{C3}$ were also fixed to one, then the constraint becomes redundant and is just dropped. If the constraint becomes a set covering constraint in a BnB node, the domain propagation techniques that are specific to set-covering constraints are applied.

A set covering constraint is one that takes the form

$$\sum_{Q_{binary}} \ddot{x}_j \geq 1. \tag{5.6}$$

This has the same propagation rule as the set packing constraints' propagation rule. However, we use a method implemented in SCIP that handles propagation for set-covering constraints differently, with a tailored more efficient way that can handle large numbers of set covering constraints using the fact that a set covering constraint is equivalent to a clause $l_1 \vee l_2 \vee \ldots \vee |l_{Q_{binary}}|$. To propagate the clause, the method of *two watched literals* is used [72], which relies on the main observation that implications are derived from clauses only if all but one literal in a clause are fixed to zero. Therefore, a clause is considered for propagation only if the number of literals fixed to zero increases from $|l_{Q_{binary}} - 2|$ to $|l_{Q_{binary}} - 1|$. The remaining unfixed literal in this case is implied and fixed to 1.

It is sufficient to watch the state of only two arbitrary literals of the constraint, as long as both remain unfixed, the constraint does not need to be processed as no propagation can be applied. If one literal is zero, the other literals of the clause are inspected. If at least one of them is fixed to 1, the constraint is then deduced to be redundant and is removed from the node's subproblem. If at least one literal in the clause is unfixed, then watching the fixed literal stops and the unfixed one gets watched instead. Finally, if all the literals except the other watched literal are fixed to zero, the other watched literal becomes implied and fixed to one.

3. **Variable Bound Constraints:** Variable bound constraints are generally defined as:

$$\underline{\beta} \le \ddot{x}_i + a_j \ddot{x}_j \le \overline{\beta} \quad i, j : i \neq j, \tag{5.7}$$

with $\ddot{x}_j \in \mathbb{Z}$, $a_j \in \mathbb{R}$ and $\underline{\beta}, \overline{\beta} \in \mathbb{R} \cup \pm\infty$. A special case of it is the variable upper bound constraint that takes the form

$$\ddot{x}_i \le u_i' \ddot{x}_j, \quad i, j : i \neq j, \tag{5.8}$$

where u_i' is the local upper bound of the variable $\ddot{x}_i$ and $\ddot{x}_j$ is binary. Constraint set $\overline{C5}$ in formulation $\mathcal{HAP}_{MBQCP}^{Eff}$ belongs to the category of variable upper bound constraints and the domain propagation of the variables in that constraint set, namely $y_{m,i,c,t}$ and $p_{m,i,c,t}$ in any of BnB nodes follow the simple rules:

(a) fix $p_{m,i,c,t}$ to zero if $y_{m,i,c,t} = 0$, and
(b) fix $y_{m,i,c,t}$ to 1 if $p_{m,i,c,t} \neq 0$.

4. **Problem $\mathcal{HAP}_{MBQCP}^{Eff}$ specific Constraint set $\overline{C1}$ Propagation:** The simple constraint set $\overline{C1}$ enables the propagation of many reductions if the binary constant $\lambda_{m,k}$ is zero for some of the constraints in the set. For these constraints, the corresponding variables $\phi_{m,k}$ are immediately deduced to be zeros and their corresponding constraints dropped. The $\overline{C1}$ constraints for the remaining variables $\phi_{m,k}$ for which the corresponding binary constant is $\lambda_{m,k} = 1$, also get dropped because they become redundant implying that the binary variables $\phi_{m,k} \le 1$. This procedure propagates further reductions in the root node and possibly other nodes. The propagations would lead to the following:

(a) All the constraints in $\overline{C3}, \overline{C8}, \overline{Q2}, \overline{Q4}$ in which $\phi_{m,k} = 0$ become redundant and can be removed.

    (b) If for any m, there are fixed variables $\phi_{m,k} = 0 \; \forall k$ in a BnB node subproblem, all the variables $y_{m,i,c,t}$ for that given m then consequently get fixed to zero due to constraint set $\overline{C4}$. This then leads to,

    (c) fixing θ_m to zero according to constraint set $\overline{C10}$.

5. **<u>Objective Function Propagation:</u>** This type of propagation infers bounds on decision variables that are valid since taking a value for the variable outside its tightened bounds will not lead to a better incumbent (primal) solution. Taking the objective function of formulation $\mathcal{HAP}_{MBQCP}^{Eff}$, which is always guaranteed to be integer, and does not include any continuous variables, then inferior solutions are ruled out using:

$$\sum_{m}^{M}\sum_{k}^{K} \rho_{m,k}\phi_{m,k} \geq \ddot{c}_{primal} - \left(1 - \hat{\delta}\right), \qquad (5.9)$$

where $\hat{\delta} \in \mathbb{R}^+$ is the optimality tolerance and $\ddot{c}_{primal}$ is the objective function value of the best solution obtained so far in the BnB algorithm. This objective constraint can be easily transformed to a binary knapsack constraint by:

- multiplying the constraint by -1,
- complementing the variables with negative coefficients by $\overline{\phi}_{m,k} = 1 - \phi_{m,k}$, and
- rounding down the resulting right hand side constant.

We use a computationally efficient knapsack propagator implemented in SCIP [62] to reduce the variable bounds and propagate further reductions for variables that belong to the objective function in $\mathcal{HAP}_{MBQCP}^{Eff}$, in this case $\phi_{m,k}, \; \forall \, m, k$ only.

5.3 Primal Heuristics

Heuristics are known to be quick in terms of obtaining solutions for an optimization problem; however, they are not guaranteed to obtain a solution, least of all an optimal solution. The branch-and-bound algorithm for solving an MBQCP requires too much computational effort that grows exponentially with the size of the problem; however, it is a guaranteed method in obtaining good feasible solutions, including the optimal solution, in a finite amount of time. Therefore, combining heuristics with the BnB algorithm is expected to improve the overall performance in solving the problem formulation $\mathcal{HAP}_{MBQCP}^{Eff}$.

There are a number of heuristics that we used in the experiments conducted for $\mathcal{HAP}_{MBQCP}^{Eff}$. The heuristics were called in the orders of their simplicity, i.e. the most computationally simple ones were called first. The heuristics phase is terminated if a feasible solution is found. Otherwise, if a heuristic algorithm could not find a feasible solution, a more computationally involved heuristic with better chances of finding feasible solutions is invoked. Furthermore, the simple heuristics are called in all nodes, while the more complex ones are called at lower frequencies as the BnB grows deeper.

In the following subsections, the types of heuristics used in the solving procedure for $\mathcal{HAP}_{MBQCP}^{Eff}$ are mentioned, and a brief explanation for each is provided. Aside from those, we also used *Simple Rounding, Rounding, Integer Shifting, Feasibility Pump, RENS, RINS* and *Crossover* heuristics for which we refer the reader to [62], [73], [74] and [75] for detailed explanation on how these heuristics work.

5.3.1 Pseudocost Diving

Pseudocost diving belongs to a group of heuristics known as *diving heuristics* [62]. Diving heuristics, have a generic framework whose main elements perform the following:

1. Choose a variable in the set $\mathcal{FRAC}$ of fractional variables of binary type and a rounding direction.
2. Fix the binary variable to the value corresponding to its rounding direction.
3. Call domain propagation to propagate the fixed variable.

4. Resolve the relaxation $\mathcal{Q}_{relax}$ after making the necessary changes that result from fixing the chosen variable.

5. If $\mathcal{Q}_{relax}$ has no feasible solution, then stop with failure. Otherwise, using the obtained solution, repeat the entire procedure all over. These repetitions can continue until $\mathcal{FRAC} = \emptyset$, or an iteration limit is reached.

Different diving heuristics differ in the way the first step is performed. In a *pseudocost diving* heuristic, the pseudocost values Ψ_j^0 and Ψ_j^1 are used for the variable selection and deciding the rounding direction. First, a decision on the rounding direction for all the variables in $\mathcal{FRAC}$ for the solution $\ddot{x}_{relax}^{opt}$ of $\mathcal{Q}_{relax}$ is obtained by performing the following steps:

1. The values of each of the variables $\ddot{x}_{relax_j}^{opt}$: $j \in \mathcal{FRAC}$ are compared with their counterpart $\ddot{x}_{relax_j}^{\mathcal{R}}$ in the root node $\mathcal{R}$. If the difference between $\ddot{x}_{relax_j}^{opt}$ and $\ddot{x}_{relax_j}^{\mathcal{R}}$ indicates the variable is being pushed to a certain direction, then that direction is selected for rounding.

2. Otherwise, if the difference does not yield a rounding decision, if the fractional part of the variable is less than 0.3, then it is rounded downwards, while if it is greater than 0.7 it gets rounded upwards.

3. If still a rounding decision is not made, then a variable in $\mathcal{FRAC}$ is rounded to 0 if $\Psi_j^0 < \Psi_j^1$ and 1 otherwise.

Finally, the variable in $\mathcal{FRAC}$ that maximizes:

$$\sqrt{1 - \ddot{x}_{relax_j}^{opt}} \cdot \frac{1 + \Psi_j^1}{1 + \Psi_j^0} \; (downwards) \quad or$$
$$\sqrt{\ddot{x}_{relax_j}^{opt}} \cdot \frac{1 + \Psi_j^0}{1 + \Psi_j^1} \; (upwards). \tag{5.10}$$

is chosen for rounding [62].

5.3.2 Clique Partition-Based Large Neighborhood Search Heuristic

Large neighborhood search (LNS) heuristics [76] restrict the search for good feasible solutions to a neighborhood of a certain reference point. This restriction makes a node $\mathcal{Q}$ subproblem easier to solve with a better chance of finding a high-quality feasible solution. The restricted subproblems do not need to be solved to optimality, but to a solution that is better than the current incumbent. The neighborhood is defined by introducing additional constraints to the node's subproblem which are usually variable fixings. The success of an LNS is strongly tied to defining a good neighborhood.

The main elements of the *clique partition based LNS* are

1. First, the clique table constructed during the presolving phase is used to fix a subset of the binary variables in $\mathcal{HAP}_{MBQCP}^{Eff}$ [77]. The only type of binary variables in the objective function is the set of variables $\phi_{m,k} \; \forall \; m, k$. If any of those within the clique was not already fixed to 1, then the one with the largest coefficient is chosen to be fixed to 1. If the binary variable is not one of the $\phi_{m,k}$ variables, the selection of which variable in the clique to fix to 1 is random since their coefficients in the objective function is zero. The rest of the variables in a clique are fixed to zero using two rounds of domain propagation. The processes iterates until all binary variables are fixed.
2. If the domain propagation in any iteration detects infeasibility, one level of backtracking to undo the previous fixing and its corresponding domain propagation fixings is performed.
3. After a sufficient number of fixations (which is determined by a preset threshold), the resulting reduced (hopefully easier) LP is solved and an attempt is made to round the resulting solution to a feasible solution using the *simple rounding* heuristic.

4. The LNS is invoked again with a neighborhood defined by the fixations obtained from the last phase and a sub-MBQCP for that neighborhood is created and solved (not necessarily to optimality). If a feasible solution is found it gets returned and the heuristic terminates.

5.3.3 Undercover Heuristic

The *undercover* heuristic is suitable for any general non-convex mixed integer nonlinear programs (MINLPs) and obtains feasible solutions by solving sub-MIPs [67]. Therefore, it is an MINLP heuristic that is suitable for the more specific MBQCP class and solves sub-MBP for obtaining feasible solutions. The previously mentioned heuristics deal with the integrality as the source of complexity to the problem while the *Undercover* heuristic considers the nonconvex constraints, which are quadratic in $\mathcal{HAP}^{Eff}_{MBQCP}$, as the source of complexity. By solving a vertex covering problem, it identifies a minimal set of variables to fix in order to linearize a quadratic constraint in $\mathcal{HAP}^{Eff}_{MBQCP}$ at the BnB node in which it is called. This is based on an observation that fixing certain variables can reduce the node's sub-problem to a sub-MIP, which in the case of $\mathcal{HAP}^{Eff}_{MBQCP}$ is a sub-MBLP.

A cover for a non-linear constraint is defined as the set of variables which, if fixed, linearizes the constraint. A cover to the entire problem linearizes all the non-linear constraints. This may also require the fixing of continuous variables which could introduce significant errors that render the problem infeasible. Therefore, the heuristic tries to minimize the number of fixed variables to obtain a large sub-MIP through obtaining minimum covers.

As in [67], the co-occurance graph consists of nodes that represent the problem's variables and edges (i, j) which are present only if the Hessian matrix of some quadratic constraint has a nonzero entry for (i, j). The minimum cover is obtained by solving a pure binary linear program for the co-occurance graph.

In the next section, we provide the details of the experiments conducted on $\mathcal{HAP}^{Eff}_{MBQCP}$ using the different techniques explained in this section for the proposed BnB based framework. We also present and discuss the obtained results that indicate the relative performances of different combinations of the proposed techniques.

5.4 Computational Experiments

We used similar simulation parameters and experimental setup as explained in Section 4.4 of 4. For the algorithmic components of the proposed BnB solution framework considered in this section, the following settings were considered in the conducted experiments:

1. node selection scheme is *best first search* with a maximum plunging depth in the BnB tree of 2 [59],
2. up to one round of presolving before starting the BnB algorithm,
3. *most infeasible branching* scheme is used,
4. a maximum number of domain propagation rounds of 1000,
5. unlimited separating cut rounds at the root node,
6. a maximum number of five separating cut rounds in the rest of the nodes,
7. a maximum number of cuts per round of 2000 at the root node,
8. a maximum number of cuts per round of 100 for the rest of the nodes, and
9. a minimum orthogonality of 0.5 for a cut to enter the LP.

For the heuristics used in the conducted experiments, the priorities of calling the heuristics as well as how often they are called in the BnB tree, are given in Table 5.1. The highest priority is given the numerical value 1, while the least is given the numerical value 10. The heuristics are called in decreasing order of their priority. The computationally simple heuristics are given the highest priority, while those that are either more complex or are *improvement heuristics* are given lower priorities. The frequency parameter defines the level depths at which the heuristic is called. For example, a frequency of 2 means that the

Table 5.1 Heuristics' settings for the conducted experiments

Heuristic	Priority Level	Frequency	Frequency Offset
Simple rounding	1	1	0
Rounding	2	1	0
Integer shifting	3	1	0
Pseudocost diving	4	1	2
RENS	5	1	1
Undercover	6	2	0
Clique	7	2	1
Feasibility pump	8	3	0
RINS	9	5	0
Crossover	10	3	3

heuristic is called for nodes that are at depths 0, 2, 4, 6, ... The frequency offset parameter defines the depth in the branching tree at which the heuristic is executed for the first time. For example, frequency of 3 and frequency offset of 2 means that the heuristic is called at the depths 2, 5, 8, ...

5.4.1 Results of the Conducted Experiments

The results illustrated in Figures 5.3, 5.4, 5.5, 5.6, 5.7, 5.8 and 5.9 compare the performance of the following for the BnB framework for $\mathcal{HAP}^{Eff}_{MBQCP}$:

1. using branching only,
2. using branching and separating cuts (branch-cut),
3. using branching and domain propagation (branch-propagate),
4. using branching, separating cuts and propagation (branch-cut-propagate), and
5. using branch-cut-propagate plus heuristics.

Figure 5.3 shows that using branch-and-cut lowers the duality gap only by approximately 5% compared to using branching only for almost equal number of nodes and LP iterations. Using branch and propagate lowers the dual gap by 33% compared to branching only at the expense of an increased number of nodes by 700% and increased

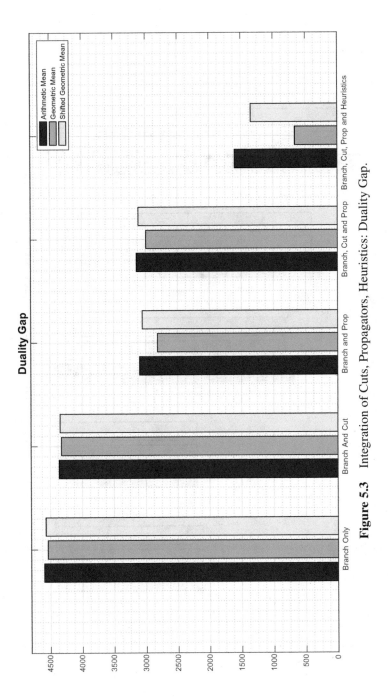

Figure 5.3 Integration of Cuts, Propagators, Heuristics: Duality Gap.

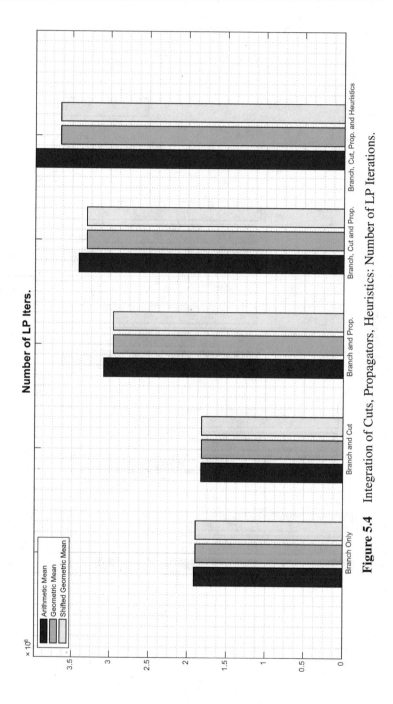

Figure 5.4 Integration of Cuts, Propagators, Heuristics: Number of LP Iterations.

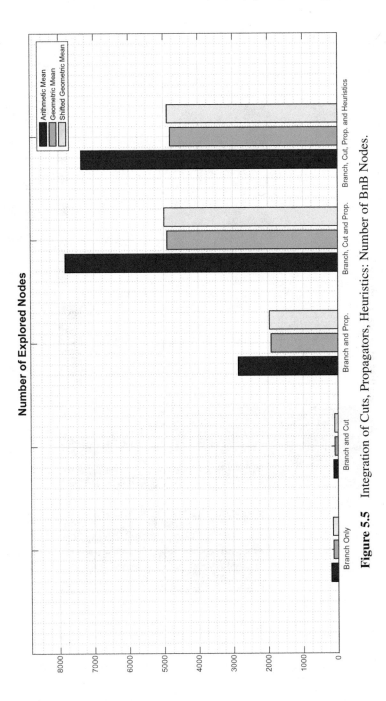

Figure 5.5 Integration of Cuts, Propagators, Heuristics: Number of BnB Nodes.

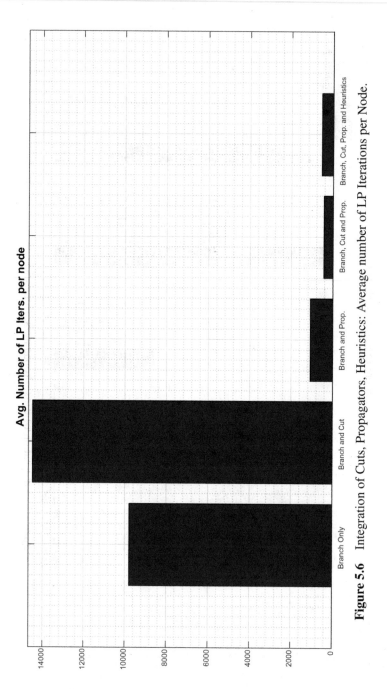

Figure 5.6 Integration of Cuts, Propagators, Heuristics: Average number of LP Iterations per Node.

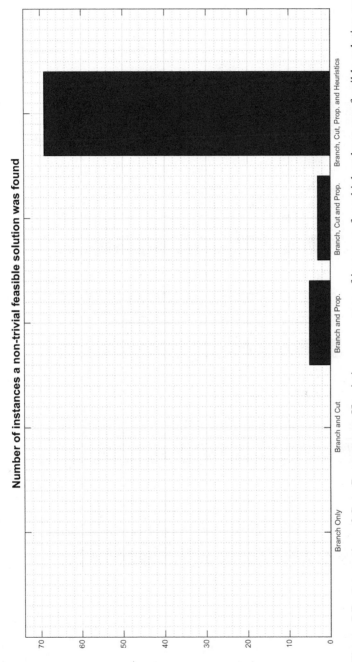

Figure 5.7 Integration of Cuts, Propagators, Heuristics: percentage of instances for which at least one feasible solution was found.

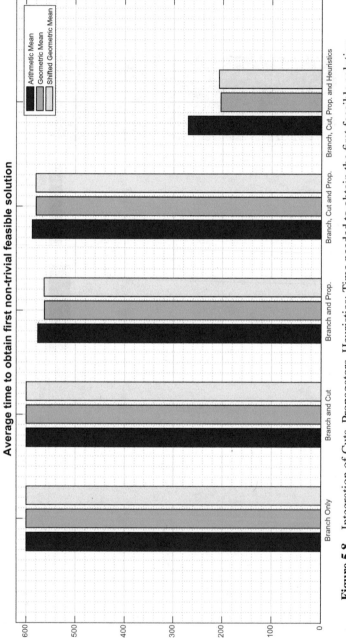

Figure 5.8 Integration of Cuts, Propagators, Heuristics: Time needed to obtain the first feasible solution.

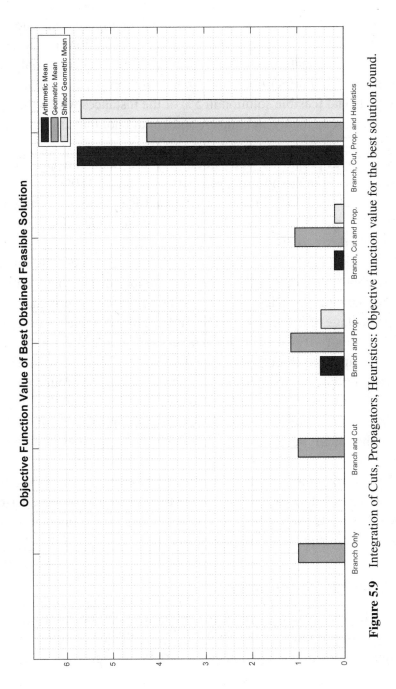

Figure 5.9 Integration of Cuts, Propagators, Heuristics: Objective function value for the best solution found.

number of LP iterations of 52% as shown in Figures 5.5 and 5.4. Moreover, Figure 5.6 shows that the average LP iterations per node for branch and propagate is much smaller than both branching only and branch-and-cut. Furthermore, the branch-and-propagate manages to get a nontrivial feasible solution in 5% of the instances within the time limit where branch only and branch-and-cut did not find any according to Figure 5.7.

Branch-cut-propagate does not lower the dual gap any further compared to branch-and-propagate. It even needs 150% more BnB nodes and around 14% higher number of LP iterations as shown in Figures 5.5 and 5.4. Moreover, the percentage of feasible solutions found goes down to 3%.

Integrating heuristics to branch-cut-propagate reduces the duality gap by about 56% due to the increased number of feasible solutions that are found (as given in Figure 5.7) which enhance the primal bound. The percentage of instances in which at least one feasible solution was found is about 70% with heuristics versus 5% and 3% for branch-propagate and branch-cut-propagate without heuristics. Also, the time needed to obtain the first feasible solution decreases by around 64% as Figure 5.8 shows. Moreover, the objective function value increases by nearly six times. This comes at the expense of an increase of 11% in the number of LP iterations compared to branch-cut-prop as shown in Figure 5.4.

We can therefore conclude that using propagators and heuristics are recommended for $\mathcal{HAP}^{Eff}_{MBQCP}$ due to the improvements in performance they yield. Separating cuts however are more computationally expensive compared to the improvements they yield for $\mathcal{HAP}^{Eff}_{MBQCP}$.

5.5 Chapter Conclusion

In this Chapter, we considered three components of the proposed BnB based solution framework for $\mathcal{HAP}^{Eff}_{MBQCP}$ which are cutting planes separation, domain propagation and a variety of primal heuristics. For each component, we described the different types used for our formulated problem for which quite are dependent on the types of constraints we have in the problem's formulation. A detailed explanation was given for the numerical experiments which were conducted to measure the performances of each of the three aspects when combined in the BnB algorithm to solve our problem. According to the results obtained from the experiments, we suggest the use of branch and propagate along with the heuristics suggested as they gave the lowest duality gap, lowest computation time to obtain the first feasible solution and had the highest percentage of instances to obtain a non-trivial feasible solution within the solving time limits. These came at the expense of a large number of explored BnB nodes and LP iterations when compared to the branching only case, where none of the three components were combined in the BnB framework.

6

Conclusion and Future Work

This chapter concludes the book and provides possible future work for extending the discussed research work.

6.1 Conclusion

This book presented part of the research work conducted by Ahmed Ibrahim towards his PhD degree. We explained a proposed extended problem (E-Prob) for an earlier problem that we considered in our earlier published works and how we were successful in obtaining a more efficient formulation (which is the major contribution of our work). For the problem considered in this book, a user was allowed to join any session being transmitted in all the neighboring cells provided the SINR threshold was met where as P-Prob, the problem we considered in our earlier works, a user only can join a subset of the sessions being transmitted in the one cell which is the one the user resides in. Moreover, E-Prob allowed a multicast group to receive transmission of a session on more than one antenna simultaneously, which is a more flexible proposal that utilizes the space dimension, that was not utilized in the primary problem. Also, the extended problem took into account heterogeneous user and session priorities, which were considered to be homogeneous in P-Prob. We succeeded in obtaining $\mathcal{HAP}_{MBQCP}^{Eff}$, a formulation that is far more efficient in terms of the size compared to formulation (OP1) in [4] used for P-Prob. The more efficient formulation $\mathcal{HAP}_{MBQCP}^{Eff}$ was a separate achievement on its

115

own that was used in E-Prob. In other words, if E-Prob was reduced to P-Prob, $\mathcal{HAP}_{MBQCP}^{Eff}$ would still be much more efficient than (OP1).

To solve the derived more efficient formulation $\mathcal{HAP}_{MBQCP}^{Eff}$ that we considered in this book, McCormick underestimators were used for linear relaxation of the bilinear terms in the nonconvex quadratic constraints. A branch-and-cut algorithm was used for solving the problem. Different constraint and objective function-specific domain propagation techniques, which are also available in the SCIP solver, were integrated. Moreover, different types of primal heuristic techniques were combined with the branch-and-cut algorithm which were shown to obtain more feasible solutions in the instances solved in shorter time. The experiments showed that domain propagation had a significant performance improvement as compared to the cutting planes used.

Furthermore, different branching techniques were compared and the results showed *cloud branching* to achieve the best balance in the trade-off between the duality gap and the computational effort. A presolving reformulation linearizion technique for a specific set of quadratic constraints in $\mathcal{HAP}_{MBQCP}^{Eff}$ was used and the experiments indicated its effect on performance for different numbers of presolving rounds. According to the results, using the technique for a number of presolving rounds that is greater than or equal to 100 improves the performance significantly.

6.2 Future Work

We believe that the presented work in this book could be extended in at least three ways. One way is using solution techniques from the optimization literature to solve the mixed binary polynomial constrained program (MBPCP) in Chapter 3 before reducing it to an MBQCP. The performances for solving both MBPCP versus solving MBQCP can be compared in terms of the goodness of the solutions, the computational effort and the memory requirements. For solving nonconvex MBPCP, the recent developments for solution techniques were achieved by Nataraj et al. in [78–80] could be considered. The approach they proposed is a Bernstein branch and prune algorithm that is based on the Bernstein polynomial approach. They introduced features like Bernstein box consistency and Bernstein hull consistency algorithms to prune the search regions. Using a Bernstein contraction algorithm, the computation of Bernstein coefficients after the pruning operation can be avoided which speeds up their algorithm. A Bernstein cutoff test based on the vertex property of the Bernstein coefficients is a key ingredient for the approach to solve a MBPCP.

Another possible extension is to use different nonlinear convex relaxation schemes for $\mathcal{HAP}^{Eff}_{MBQCP}$ aside from the linear relaxation, perhaps in a successive fashion like in [81]. These could include different combinations of second-order cone programming relaxations, semidefinite program relaxations or rank-2 valid linear inequalities [82, 83]. Furthermore, the MIQCP disjunctive cuts by Saxena et al. can be used to improve the relaxations further [84]. The two extension directions and the work presented in this book for solving $\mathcal{HAP}^{Eff}_{MBQCP}$ could then be compared through extensive experiments to find which of all these techniques would yield the tightest bounds.

The second way the work presented in this book could be extended is by relaxing the assumption that the measured channel gain values known to the HAP are ideally accurate. In a practical problem, measurement errors and prediction errors are encountered. Since the channel gain values $g_{i,k,c,t}$ constitute the problem formulation's parameters, it is important to know whether the solution obtained would be significantly different than an error free (ideal) situation, and whether it would still remain feasible to the formulation in the first place. It would be important to understand the tolerable ranges of errors for which the solution would not change and/or remain feasible, where large ranges are obviously desirable. Using *sensitivity analysis*, we can evaluate these *insensitive* ranges and learn about the sensitivity of the problem's solution to channel measurements and estimation errors outside the calculated ranges. The sensitivity of the problem's solution would essentially be the rate of the change in the objective value to the errors outside the calculated insensitive ranges. This type of knowledge could be used to determine the robustness of the optimization problem's solution and whether the channel estimation scheme is suitable for our multicasting AC-RRA problem. The observations could then be used to decide whether the channel estimation scheme would need improvement to achieve higher estimation accuracy, which would then increase the hardware complexity and cost.

The third way we can extend our work is by considering multi-rate multicast transmissions. Since in this book we proposed variants of the LCG single rate multicasting, multi-rate transmissions would be a reasonable extension to the type of multicasting used in both system models we considered. Multi-rate multicasting exploits the multi-user multi-channel diversity within a multicast group enabling users to receive heterogeneous qualities for the information transmitted to each group. This is very useful when the information transmitted is of graphical multimedia content, like video for example. Each user in the group could then perceive different video resolutions depending on its channel and interference conditions. For that purpose, *Information Decomposition Techniques* (IDT), where mutlimedia content is split to multiple sub-streams, can be a good candidate for a multi-rate transmission scheme. Every user would receive a subset of the sub-streams, which correspond to a particular video resolution.

Bibliography

[1] A. Ibrahim, *Admission control and radio resource allocation for multicasting over high altitude platforms*. PhD thesis, University of Manitoba, October 2016.

[2] A. Ibrahim and A. S. Alfa, "Radio Resource Allocation for Multicast Transmissions over High Altitude Platforms," in *IEEE Globecom Workshops (GC Wkshps), 2013*, pp. 281–287, IEEE, 2013.

[3] A. Ibrahim and A. S. Alfa, "Solving Binary and Continuous Knapsack Problems for Radio Resource Allocation over High Altitude Platforms," in *Wireless Telecommunications Symposium (WTS), 2014*, pp. 1–7, IEEE, 2014.

[4] A. Ibrahim and A. S. Alfa, "Using Lagrangian Relaxation for Radio Resource Allocation in High Altitude Platforms," *IEEE Transactions on Wireless Communications*, vol. 14, no. 10, pp. 5823–5835, 2015.

[5] E. Falletti, M. Laddomada, M. Mondin, and F. Sellone, "Integrated Services from High-Altitude Platforms: a Flexible Communication System," *IEEE Communications Magazine*, vol. 44, no. 2, pp. 85–94, 2006.

[6] G. M. Djuknic, J. Freidenfelds, and Y. Okunev, "Establishing Wireless Communications Services via High-Altitude Aeronautical Platforms: A Concept Whose Time Has Come?," *IEEE Communications Magazine*, vol. 35, no. 9, pp. 128–135, 1997.

[7] A. Mohammed, A. Mehmood, F. Pavlidou, and M. Mohorcic, "The Role of High-Altitude Platforms (HAPs) in the Global

Wireless Connectivity," *Proceedings of the IEEE*, vol. 99, no. 11, pp. 1939–1953, 2011.

[8] T. Tozer and D. Grace, "High-Altitude Platforms for Wireless Communications," *Electronics & Communication Engineering Journal*, vol. 13, no. 3, pp. 127–137, 2001.

[9] ITU-R, "Preferred Characteristics of Systems in the Fixed Service Using High Altitude Platforms Operating in the Bands 47.2–47.5 GHz and 47.9–48.2 GHz," Recommendation F.1500, International Telecommunication Union, Geneva, May 2000.

[10] ITU-R, "Technical and Operational Characteristics for the Fixed Service Using High Altitude Platform Stations in the Bands 27.5–28.35 GHz and 31–31.3 GHz," Recommendation F.1569, International Telecommunication Union, Geneva, May 2002.

[11] ITU-R, "Minimum Performance Characteristics and Operational Conditions for High Altitude Platform Stations Providing IMT-2000 in the Bands 1,885–1,980 MHz, 2,010–2,025 MHz and 2,110–2,170 MHz in Regions 1 and 3 and 1,885–1,980 MHz and 2,110–2,160 MHz in Region 2," Recommendation M.1456, International Telecommunication Union, Geneva, May 2000.

[12] ITU-R, "Technical and Operational Characteristics of Gateway Links in the Fixed Service Using High Altitude Platform Stations in the Band 5,850–7,075 MHz to Be Used in Sharing Studies," Recommendation F.1891, International Telecommunication Union, Geneva, May 2011.

[13] P. Sudheesh, M. Mozaffari, M. Magarini, W. Saad, and P. Muthuchidambaranathan, "Sum-rate analysis for high altitude platform (hap) drones with tethered balloon relay," *IEEE Communications Letters*, vol. 22, no. 6, 2018.

[14] D. Xu, X. Yi, Z. Chen, C. Li, C. Zhang, and B. Xia, "Coverage ratio optimization for hap communications," in *Personal, Indoor, and Mobile Radio Communications (PIMRC), 2017 IEEE 28th Annual International Symposium on*, pp. 1–5, IEEE, 2017.

[15] F. Dong, H. Han, X. Gong, J. Wang, and H. Li, "A constellation design methodology based on qos and user demand in high-altitude platform broadband networks," *IEEE Transactions on Multimedia*, vol. 18, no. 12, pp. 2384–2397, 2016.

[16] J. Zhang, Y. Zeng, and R. Zhang, "Spectrum and energy efficiency maximization in uav-enabled mobile relaying," in *Communications (ICC), 2017 IEEE International Conference on*, pp. 1–6, IEEE, 2017.

[17] D. Grace and M. Mohorcic, *Broadband Communications via High-Altitude Platforms*. John Wiley & Sons, 2011.

[18] T. Tozer, G. Olmo, and D. Grace, "The European HeliNet Project," 2000.

[19] F. Dovis, L. L. Presti, E. Magli, G. Olmo, and F. Sellone, "HeliNet: A Network of UAV-HAVE Stratospheric Platforms. System Concepts and Applications to Environmental Surveillance," in *Data Systems in Aerospace*, vol. 457, p. 551, 2000.

[20] D. Grace, M. Mohorcic, M. Oodo, M. Capstick, M. B. Pallavicini, and M. Lalovic, "CAPANINA-Communications from Aerial Platform Networks Delivering Broadband Information for All," *Proceedings of the 14th IST Mobile and Wireless and Communications Summit*, 2005.

[21] " CAPANINA project for the development of broadband communications capability HAPs." http://www.capanina.org.

[22] D. Grace, M. H. Capstick, M. Mohorcic, J. Horwath, M. B. Pallavicini, and M. Fitch, "Integrating Users into the Wider Broadband Network via High Altitude Platforms," *IEEE Wireless Communications*, vol. 12, no. 5, pp. 98–105, 2005.

[23] R. v. Nee and R. Prasad, *OFDM for Wireless Multimedia Communications*. Artech House, Inc., 2000.

[24] J. Thornton, D. Grace, M. H. Capstick, and T. C. Tozer, "Optimizing an Array of Antennas for Cellular Coverage from a High Altitude Platform," *IEEE Transactions on Wireless Communications*, vol. 2, no. 3, pp. 484–492, 2003.

[25] L. Zhao, L. Cong, F. Liu, K. Yang, and H. Zhang, "Joint Time-Frequency-Power Resource Allocation for Low-Medium-Altitude Platforms-Based WiMAX Networks," *IET communications*, vol. 5, no. 7, pp. 967–974, 2011.

[26] A. Abrardo and D. Sennati, "Centralized Radio Resource Management Strategies with Heterogeneous Traffics in HAPS WCDMA Cellular Systems," *IEICE Transactions on Communications*, vol. 86, no. 3, pp. 1040–1049, 2003.

[27] Y. C. Foo, W. L. Lim, and R. Tafazolli, "Centralized Total Received Power Based Call Admission Control for High Altitude Platform Station UMTS," in *The 13th IEEE International Symposium on Personal, Indoor and Mobile Radio Communications, 2002.*, vol. 4, pp. 1596–1600, IEEE, 2002.

[28] Y. C. Foo, W. L. Lim, and R. Tafazolli, "Centralized Downlink Call Admission Control for High Altitude Platform Station UMTS with Onboard Power Resource Sharing," in *Vehicular Technology Conference, 2002. Proceedings. VTC 2002-Fall. 2002 IEEE 56th*, vol. 1, pp. 549–553, IEEE, 2002.

[29] Y. C. Foo, W. L. Lim, and R. Tafazolli, "Call Admission Control Schemes for High Altitude Platform Station and Terrestrial Tower-Based Hierarchical UMTS," in *The Ninth International Conference on Communications Systems, 2004. ICCS 2004.*, pp. 531–536, IEEE, 2004.

[30] Y. C. Foo and W. L. Lim, "Speed and Direction Adaptive Call Admission Control for High Altitude Platform Station (HAPS) UMTS," in *Military Communications Conference, 2005. MILCOM 2005. IEEE*, pp. 2182–2188, IEEE, 2005.

[31] G. Araniti, A. Molinaro, and A. Iera, "Multicast in Terrestrial-HAP Systems," in *Vehicular Technology Conference, 2007. VTC-2007 Fall. 2007 IEEE 66th*, pp. 154–158, IEEE, 2007.

[32] Y. Liu, D. Grace, and P. D. Mitchell, "Exploiting Platform Diversity for GoS Improvement for Users with Different High Altitude Platform Availability," *IEEE Transactions on Wireless Communications*, vol. 8, no. 1, pp. 196–203, 2009.

[33] L. Zhao, J. Yi, F. Adachi, C. Zhang, and H. Zhang, "Radio Resource Allocation for Low-Medium-Altitude Aerial Platform Based TD-LTE Networks against Disaster," in *2012 IEEE 75th, Vehicular Technology Conference (VTC Spring)*, pp. 1–5, May 2012.

[34] L. Zhao, C. Zhang, H. Zhang, X. Li, and L. Hanzo, "Power-Efficient Radio Resource Allocation for Low-Medium-Altitude Aerial Platform Based TD-LTE Networks," in *IEEE Vehicular Technology Conference (VTC)*, pp. 1–5, IEEE, 2012.

[35] P. Agashe, R. Rezaiifar, and P. Bender, "CDMA2000® High Rate Broadcast Packet Data Air Interface Design," *IEEE Communications Magazine*, vol. 42, no. 2, pp. 83–89, 2004.

[36] J. Xu, S.-J. Lee, W.-S. Kang, and J.-S. Seo, "Adaptive Resource Allocation for MIMO-OFDM Based Wireless Multicast Systems," *IEEE Transactions on Broadcasting*, vol. 56, no. 1, pp. 98–102, 2010.

[37] K. Bakanoglu, W. Mingquan, L. Hang, and M. Saurabh, "Adaptive Resource Allocation in Multicast OFDMA Systems," in *Wireless Communications and Networking Conference (WCNC), 2010 IEEE*, pp. 1–6, IEEE, 2010.

[38] J. Liu, W. Chen, Z. Cao, and K. B. Letaief, "Dynamic Power and Sub-carrier Allocation for OFDMA-Based Wireless Multicast Systems," in *IEEE International Conference on Communications, 2008. ICC'08.*, pp. 2607–2611, IEEE, 2008.

[39] P. K. Gopala and H. El Gamal, "On the Throughput-Delay Trade-off in Cellular Multicast," in *International Conference on Wireless Networks, Communications and Mobile Computing, 2005*, vol. 2, pp. 1401–1406, IEEE, 2005.

[40] H. Won, H. Cai, K. Guo, A. Netravali, I. Rhee, K. Sabnani, *et al.*, "Multicast Scheduling in Cellular Data Networks," *IEEE Transactions on Wireless Communications*, vol. 8, no. 9, pp. 4540–4549, 2009.

[41] C. Suh and J. Mo, "Resource Allocation for Multicast Services in Multicarrier Wireless Communications," *IEEE Transactions on Wireless Communications*, vol. 7, pp. 27–31, Jan 2008.

[42] C. Suh and J. Mo, "Resource Allocation for Multicast Services in Multicarrier Wireless Communications," in *INFOCOM 2006. 25th IEEE International Conference on Computer Communications. Proceedings*, pp. 1–12, April 2006.

[43] P. Eusiebio and A. Correia, "Two QoS Regions Packet Scheduling for MBMS," in *2nd International Symposium on Wireless Communication Systems, 2005.*, pp. 777–781, Sept 2005.

[44] S. Deb, S. Jaiswal, and K. Nagaraj, "Real-Time Video Multicast in WiMAX Networks," in *INFOCOM 2008. The 27th Conference on Computer Communications. IEEE*, April 2008.

[45] M. Shao, S. Dumitrescu, and X. Wu, "Layered Multicast With Inter-Layer Network Coding for Multimedia Streaming," *IEEE Transactions on Multimedia*, vol. 13, pp. 353–365, April 2011.

[46] C. H. Koh and Y. Y. Kim, "A Proportional Fair Scheduling for Multicast Services in Wireless Cellular Networks," in *Vehicular Technology Conference, 2006. VTC-2006 Fall. 2006 IEEE 64th*, pp. 1–5, Sept 2006.

[47] T. Han and N. Ansari, "Energy Efficient Wireless Multicasting," *IEEE Communications Letters*, vol. 15, no. 6, pp. 620–622, 2011.

[48] N. Shrestha, P. Saengudomlert, and Y. Ji, "Dynamic Subcarrier Allocation with Transmit Diversity for OFDMA-Based Wireless Multicast Transmissions," in *2010 International Conference on Electrical Engineering/Electronics Computer Telecommunications and Information Technology (ECTI-CON),*, pp. 410–414, May 2010.

[49] S. M. Elrabiei and M. H. Habaebi, "Energy Efficient Cooperative Multicasting for MBS WiMAX Traffic," in *2010 5th IEEE International Symposium on Wireless Pervasive Computing (ISWPC)*, pp. 600–605, May 2010.

[50] N. Sharma and A. S. Madhukumar, "Genetic Algorithm Aided Proportional Fair Resource Allocation in Multicast OFDM Systems," *IEEE Transactions on Broadcasting*, vol. 61, pp. 16–29, March 2015.

[51] H. Won, H. Cai, D. Y. Eun, K. Guo, A. Netravali, I. Rhee, and K. Sabnani, "Multicast Scheduling in Cellular Data Networks," in *INFOCOM 2007. 26th IEEE International Conference on Computer Communications. IEEE*, pp. 1172–1180, May 2007.

[52] B. Da, C. C. Ko, and Y. Liang, "An Enhanced Capacity and Fairness Scheme for MIMO-OFDMA Downlink Resource Allocation," in *International Symposium on Communications and Information Technologies, 2007. ISCIT '07.*, pp. 495–499, Oct 2007.

[53] R. O. Afolabi, A. Dadlani, and K. Kim, "Multicast Scheduling and Resource Allocation Algorithms for OFDMA-Based Systems: A survey," *IEEE Communications Surveys & Tutorials*, vol. 15, no. 1, pp. 240–254, 2013.

[54] G. P. McCormick, "Computability of Global Solutions to Factorable Nonconvex Programs: Part I–Convex Underestimating Problems," *Mathematical programming*, vol. 10, no. 1, pp. 147–175, 1976.

[55] E. L. Lawler and D. E. Wood, "Branch-and-Bound Methods: A Survey," *Operations research*, vol. 14, no. 4, pp. 699–719, 1966.

[56] T. Berthold, A. M. Gleixner, S. Heinz, and S. Vigerske, "Analyzing the Computational Impact of MIQCP Solver Components," *Numerical Algebra, Control and Optimization*, vol. 2, no. 4, pp. 739–748, 2012.

[57] T. Berthold, S. Heinz, and S. Vigerske, *Extending a CIP Framework to Solve MIQCPs*. Springer, 2012.

[58] D. A. Pearce and D. Grace, "Optimum Antenna Configurations for Millimetre-Wave Communications from High-Altitude Platforms," *IET Communications*, vol. 1, no. 3, pp. 359–364, 2007.

[59] T. Achterberg, "SCIP: Solving Constraint Integer Programs," *Mathematical Programming Computation*, vol. 1, no. 1, pp. 1–41, 2009.

[60] S. Burer and A. Saxena, "The MILP Road to MIQCP," in *Mixed Integer Nonlinear Programming*, pp. 373–405, Springer, 2012.

[61] W. L. Winston, M. Venkataramanan, and J. B. Goldberg, *Introduction to Mathematical Programming*, vol. 1. Thomson/Brooks/Cole, 2003.

[62] T. Achterberg, *Constraint Integer Programming*. PhD thesis, Technische Universität Berlin, 2007.

[63] T. Achterberg, "Solving Constraint Integer Programs (SCIP) Solver Documentation." http://scip.zib.de/doc/html_devel/branch_relpscost_8c_source.php.

[64] T. Berthold and D. Salvagnin, "Cloud Branching," in *Integration of AI and OR Techniques in Constraint Programming for Combinatorial Optimization Problems*, pp. 28–43, Springer, 2013.

[65] J. Currie and D. I. Wilson, "OPTI: Lowering the Barrier Between Open Source Optimizers and the Industrial MATLAB User," in *Foundations of Computer-Aided Process Operations* (N. Sahinidis and J. Pinto, eds.), (Savannah, Georgia, USA), 8–11 January 2012.

[66] MathWorks, "Parallel Computing Toolbox User's Guide." http://www.mathworks.com/help/distcomp/index.html.

[67] T. Berthold and A. M. Gleixner, "Undercover: A Primal MINLP Heuristic Exploring a Largest sub-MIP," *Mathematical Programming*, vol. 144, no. 1–2, pp. 315–346, 2014.

[68] E. Balas, S. Ceria, G. Cornuéjols, and N. Natraj, "Gomory Cuts Revisited," *Operations Research Letters*, vol. 19, no. 1, pp. 1–9, 1996.

[69] M. W. Savelsbergh, "Preprocessing and Probing Techniques for Mixed Integer Programming Problems," *ORSA Journal on Computing*, vol. 6, no. 4, pp. 445–454, 1994.

[70] R. Borndörfer and Z. Kormos, "An Algorithm for Maximum Cliques," *Unpublished working paper, Konrad-Zuse-Zentrum für Informationstechnik Berlin*, 1997.

[71] F. Domes and A. Neumaier, "Constraint Propagation on Quadratic Constraints," *Constraints*, vol. 15, no. 3, pp. 404–429, 2010.

[72] M. W. Moskewicz, C. F. Madigan, Y. Zhao, L. Zhang, and S. Malik, "Chaff: Engineering an Efficient SAT Solver," in *Proceedings of the 38th Annual Design Automation Conference*, pp. 530–535, ACM, 2001.

[73] T. Berthold, "RENS: The Optimal Rounding," Tech. Rep. 12–17, ZIB, Takustr.7, 14195 Berlin, 2012.

[74] T. Achterberg and T. Berthold, "Improving the Feasibility Pump," *Discrete Optimization*, vol. 4, no. 1, pp. 77–86, 2007.

[75] T. Berthold, "Primal Heuristics for Mixed Integer Programs," 2006.

[76] T. Berthold, S. Heinz, M. E. Pfetsch, and S. Vigerske, "Large Neighborhood Search Beyond MIP," in *Proceedings of the 9th Metaheuristics International Conference (MIC 2011)*, pp. 51–60, 2011.

[77] G. Gamrath, T. Berthold, S. Heinz, and M. Winkler, "Structure-Based Primal Heuristics for Mixed Integer Programming," in *Optimization in the Real World*, vol. 13, pp. 37–53, 2015.

[78] P. Nataraj and M. Arounassalame, "Constrained Global Optimization of Multivariate Polynomials using Bernstein Branch and Prune Algorithm," *Journal of Global Optimization*, vol. 49, no. 2, pp. 185–212, 2011.

[79] B. V. Patil, P. S. Nataraj, and S. Bhartiya, "Global Optimization of Mixed-Integer Nonlinear (Polynomial) Programming Problems: the Bernstein Polynomial Approach," *Computing*, vol. 94, no. 2–4, pp. 325–343, 2012.

[80] P. Nataraj and M. Arounassalame, "A New Subdivision Algorithm For the Bernstein Polynomial Approach to Global Optimization," *International Journal of Automation and Computing*, vol. 4, no. 4, pp. 342–352, 2007.

[81] M. Kojima and L. Tunçel, "Cones of Matrices and Successive Convex Relaxations of Nonconvex Sets," *SIAM Journal on Optimization*, vol. 10, no. 3, pp. 750–778, 2000.

[82] S. Kim and M. Kojima, "Second Order Cone Programming Relaxation of Nonconvex Quadratic Optimization Problems," *Optimization Methods and Software*, vol. 15, no. 3–4, pp. 201–224, 2001.

[83] S. Kim and M. Kojima, "Exact Solutions of Some Nonconvex Quadratic Optimization Problems via SDP and SOCP Relaxations," *Computational Optimization and Applications*, vol. 26, no. 2, pp. 143–154, 2003.

[84] A. Saxena, P. Bonami, and J. Lee, "Convex Relaxations of Non-Convex mixed Integer Quadratically Constrained Programs: Extended Formulations," *Mathematical Programming*, vol. 124, no. 1–2, pp. 383–411, 2010.

Index

About the Authors

Dr. Ahmed Ibrahim is currently a postdoctoral fellow and a per course instructor with the Memorial University of Newfoundland. He obtained his Ph.D. in Electrical and Computer Engineering from the University of Manitoba (UoM) in Electrical and Computer Engineering in July 2016 and was the recipient of the IGSES and IGSS scholarships in UoM. His research area covers, but is not limited to radio resource allocation, cross layer design and optimization, device-to-device communications, link adaptation, heterogeneous networks, backhauling using millimeter technology, network performance analysis, UAV HAP communications, wireless sensor networks and scheduling for video streaming. Dr. Ibrahim teaches courses in communication networks and wireless communications. He also serves as a reviewer and TPC member in a number of IEEE journals and conferences like TWC, communication letters, Systems Journal, IEEE Access, JSTSP, Globecom, ICC, VTC and COMNETSAT.

Attahiru S. Alfa is Professor Emeritus at the University of Manitoba, Department of Electrical and Computer Engineering and also a UP/CSIR co-hosted SARChI Chair Professor at the University of Pretoria, Department of Electrical, Electronic and Computer Engineering. Dr. Alfa's most recent research focus covers wireless sensor networks, cognitive radio networks, network restoration tools for wireless sensor networks, and the role of 5G on IoT, with specific interest in the mathematical modeling of those systems. His general research covers, but not limited to, the following areas: queueing theory and applications, optimization, performance analysis and resource allocation in telecommunication systems, modeling of communication networks, analysis of cognitive radio networks, modeling and analysis of wireless sensor networks, and smart cities. Some of his previous works include developing efficient decoding algorithms for LDPC codes, channel modeling, traffic estimation for the Internet, and cross layer analysis. Dr. Alfa also works in the application of queueing theory to other areas such as transportation systems, manufacturing systems and healthcare systems. He has authored two books, "Queueing Theory for Telecommunications: Discrete Time Modelling of a Single Node System", published by Springer in 2010, and "Applied Discrete-Time Queue" published in 2015, also by Springer, as a second edition of the first book.